THERMAL ANALYSIS: COMPARATIVE STUDIES ON MATERIALS

THERMAL ANALYSIS: COMPARATIVE STUDIES ON MATERIALS

Edited by

H. Kambe, University of Tokyo, Tokyo, Japan
P. D. Garn, University of Akron, Akron, Ohio, U.S.A.

*Proceedings of the U.S.-Japan Joint Seminar
held in Akron, Ohio on April 8–12, 1974*

1974

A HALSTED PRESS BOOK
KODANSHA LTD.
Tokyo

JOHN WILEY & SONS
New York-London-Sydney-Toronto

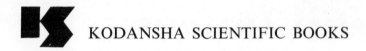

KODANSHA SCIENTIFIC BOOKS

Copyright © 1974 by Kodansha Ltd.

Library of Congress Cataloging in Publication Data
U.S.-Japan Joint Seminar, Akron, Ohio, 1974.
Thermal analysis.: comparative studies on materials.
(Kodansha Scientific books)
"A Halsted Press book."
1. Thermal analysis--Congresses. I. Kambe, Hirotaro, 1920- ed. II. Garn, Paul D., ed.
III. Title.
QD79.T38U54 1974 543'.086 74-11511
ISBN 0-470-45567-5

Published in Japan by
KODANSHA LTD.
12-21 Otowa 2-chome, Bunkyo-ku, Tokyo 112, Japan
Published by
HALSTED PRESS
a Division of John Wiley & Sons, Inc.
605 Third Avenue, New York, N.Y. 10016, U.S.A.

PRINTED IN JAPAN

CONTRIBUTORS

G.T. Armstrong   Physical Chemistry Div., Institute for Materials Research, National Bureau of Standards, Washington, D.C., U.S.A.

E.M. Barrall II   IBM Research Laboratory, San Jose, Calif., U.S.A.

R.N. Boyd   Physical Chemistry Div., Institute for Materials Research, National Bureau of Standards, Washington, D.C., U.S.A.

H. L. Friedman   General Electric Co., Re-Entry and Environmental Systems Div., Philadelphia, Pennsylvania, U.S.A.

P.K. Gallagher   Bell Laboratories, Murry Hill, New Jersey, U.S.A.

P.D. Garn   The Univ. of Akron, Ohio, U.S.A.

R.N. Goldberg   Physical Chemistry Div., Institute for Materials Research, National Bureau of Standards, Washington, D.C., U.S.A.

W.R. Griffin   Air Force Materials Laboratory, Wright-Patterson Air Force Base, Ohio, U.S.A.

K. Hirano   Dept. of Materials Science, Faculty of Engineering, Tohoku Univ., Sendai, Japan

H.W. Hoyer   Hunter College of the City Univ. of New York, New York, U.S.A.

N. Imai   Dept. of Mineral Industry, Waseda Univ., Tokyo, Japan

T. Ishii   Div. of Applied Chemistry, Faculty of Engineering, Hokkaido Univ., Sapporo, Japan

H. Kambe   Institute of Space and Aeronautical Science, Univ. of Tokyo, Meguro-ku, Tokyo, Japan

H. Kanetsuna   Research Institute for Polymers and Textiles, Yokohama, Japan

T. Kato   Institute of Space and Aeronautical Science, Univ. of Tokyo, Meguro-ku, Tokyo, Japan

M. Kochi   Institute of Space and Aeronautical Science, Univ. of Tokyo, Meguro-ku, Tokyo, Japan

R.J. Krzewki	Polymer Science and Engineering, Univ. of Massachusetts, Amherst, Mass., U.S.A.
Y. Maeda	Research Institute for Polymers and Textiles, Yokohama, Japan
J.J. Maurer	Corporate Research Laboratories, Exxon Research and Engineering Co., New Jersey, U.S.A.
M. Murakami	Institute of Space and Aeronautical Science, Univ. of Tokyo, Meguro-ku, Tokyo, Japan
R. Otsuka	Dept. of Mineral Industry, Waseda Univ., Tokyo, Japan
T. Ozawa	Electrotechnical Laboratory, Tanashi, Tokyo, Japan
R.S. Porter	Polymer Science and Engineering, Univ. of Massachusetts, Amherst, Mass., U.S.A.
E.J. Prosen	Physical Chemistry Div., Institute for Materials Research, National Bureau of Standards, Washington, D.C., U.S.A.
T. Sakamoto	Dept. of Mineral Industry, Waseda Univ., Tokyo, Japan
S. Seki	Dept. of Chemistry, Faculty of Science, Osaka Univ., Osaka, Japan
B.R. Staples	Physical Chemistry Div., Institute for Materials, Research, National Bureau of Standards, Washington, D.C., U.S.A.
H. Suga	Dept. of Chemistry, Faculty of Science, Osaka Univ., Osaka, Japan
Y. Takahashi	Dept. of Nuclear Engineering, Univ. of Tokyo, Bunkyo-ku, Tokyo, Japan

PREFACE

The general field of thermal analysis has grown rapidly in several fields almost totally independently. Substantially different practices have developed by reason of particular utility in the individual fields. This state of affairs has developed largely because of the wide applicability of the several techniques.

Advances are generally made by an individual scientist and applied first to the field in which he is working. The compartmentalization of scientific meetings tends to inhibit useful exchange of information. The need to keep abreast of his general field impels attendance at those meeting as compared to a meeting devoted to specialized techniques. Even meetings which are technique oriented, such as the triennial International Conferences on Thermal Analysis, do not enable intensive discussion because of time limitations.

The number of practitioners of thermal analysis, judging from the sale of instruments, is in the many thousands, enough that dissemination of knowledge is technologically and scientifically fruitful, but too diverse in interests to maintain good communication within the somewhat narrow range of the useful application of specialized techniques to their general fields. The most effective way of stimulating interchange of knowledge is by discourse amongst the experts, providing each an opportunity to probe for information he does not have and exposing him to similar probing from others. The task of sorting out the transferable knowledge from another field to his own is in part an exercise in imagination but reasoned judgements can be best by those very familiar with their own fields. The transfer of knowledge can be expected to improve or stimulate research and writing by an individual so that the knowledge is soon transmitted to them who recognize that individual as an authority. The discourse is best obtained in a meeting large enough to be comprehensive in character without losing the opportunity for individual contact.

For these reasons, a Joint U.S.-Japan Seminar, sponsored by the U.S. National Science Foundation and the Japan Society for the Promotion of Science, was held at The University of Akron on 8-11 April 1974. The meeting was organized by these editors.

While the general theme was the comparative investigations of new materials, some more general discussions, particularly concerning recent developments in non-isothermal kinetics, was included along with some reports on new studies on old materials. A very few topics which might appear as glaring omissions are not included because schedule conflicts prevented the invited speakers from attending.

In addition to the scientific lectures and discussions, the practical needs for aids in communicating were taken into account. The then-chairman of ICTA's Committee on Standardization, Dr. H.G. McAdie, discussed standard reference materials and requested that the participants let him know of anticipated needs as well as materials for present programs. Dr. H.G. Wiedemann, now vice chairman of that committee, also attended.

The format for the scientific program comprised an oral presentation of the essential points in the previously circulated manuscript followed by a discussion which was not formally limited in point of time. Each participant had also reviewed two other (assigned) manuscripts and the prepared questions were given to the author in advance.

The papers appearing here are the manuscripts prepared for the seminar, edited in some cases in response to questions at the seminar or to keep within a reasonable space allocation.

The editors are especially grateful to The University of Akron for providing the meeting place, its welcome to the participants and observers and the extensive staff work in assisting in the planning.

On behalf of the participants, we also thank our several organizations for the opportunity to attend and for the assistance in preparing manuscripts, and the U.S. National Science Foundation and the Japan Society for the Promotion of Science for their sponsorship of the seminar.

August 1974

Hirotaro Kambe
Paul D. Garn

CONTENTS

IX

LIQUID CRYSTALLINE BEHAVIOR OF PALMITATE ESTERS
OF SOME STEROLS STRUCTURALLY RELATED TO CHOLESTEROL

Rudolf J. Krzewki[*] and Roger S. Porter
Polymer Science and Engineering
University of Massachusetts
Amherst, Massachusetts
U.S.A.

ABSTRACT

Temperatures and heats of transition have been determined by differential scanning calorimetry for palmitate esters of campesterol, sitosterol and stigmasterol. This behavior is compared with cholesteryl palmitate as a reference compound. Structural changes in the end chain of cholesterol are thus found to affect the formation of mesophases. Whereas cholesteryl palmitate exhibits both smectic and cholesteric mesophases, the only mesophase observed with palmitate esters of the three related sterols was a monotropic smectic mesophase of campesteryl ester. Mesomorphic behavior was also investigated for palmitate esters of lophenol and 7-dehydrocholesterol. Both esters form enantiotropic smectic and cholesteric mesophases, but the latter is stable only with a narrow temperature interval. Results indicate that the presence of Δ^7 double bond does not prevent formation of a cholesteric mesophase, as was thought, but only restricts its stability to a narrow temperature range.

INTRODUCTION

The effect of the aliphatic chain length on the temperatures and heats of transition for cholesteryl esters has been studied extensively for compounds having up to 20 carbon atoms in the ester chain. The lower homologues exhibit only a cholesteric mesophase; both cholesteric and smectic mesophases are formed by the longer chain esters (1-4).

Corresponding derivatives of other sterols do not necessarily even show mesomorphic behavior. There has been recent interest in the preparation of esters of structurally-related sterols in the effort to investigate the effect of structural modification on mesomorphic properties of these esters.

Very closely similar in their structure to cholesterol are three plant sterols - campesterol, sitosterol and stigmasterol. Campesterol differs from cholesterol by having a methyl group on the C-24. Sitosterol has an ethyl group on the C-24 and stigmasterol has, in addition, a double bond in the position

---

[*]Present Address: T. R. Evans Research Center, Diamond Shamrock Corporation, Painesville, Ohio.

1

C-22 (23). These three palmitate sterol esters were thus chosen for calorimetric investigation since they represent a convenient way of studying the effect of structural changes in the sterol end chain upon the mesomorphic properties at fixed fatty acid chain length and an identical sterane skeleton.

In order to investigate the effect of a nucleus unsaturation on the liquid-crystalline behavior, palmitate esters of 7-dehydrocholesterol and lophenol are included. Both sterols have Δ^7 double bond whereas the sterol end chain is the same as cholesterol. The structures of the sterols studied as palmitate esters are given in Figure 1.

There are no literature values for temperature and heats of transition for campesteryl palmitate. Only a single solid-isotropic transition, 83.5-85.5 (5), 85.5 (6) and 91-94°C (7), is given for sitosteryl palmitate with the values in distinct disagreement. With similar uncertainty, the solid-isotropic transition for stigmasteryl palmitate is reported at 95-96 (8), 99 (9), 99.5 (6) and 101-102°C (7). A monotropic smectic-isotropic transition has also been reported for this compound (8). No transition heats were determined in any of these investigations.

Palmitate esters of 7-dehydrocholesterol and lophenol have also been investigated for mesomorphic behavior but only by optical means (10,11). Both were found to form only an enantiotropic smectic mesophase; no indications of a cholesteric mesophase were observed. The solid-smectic transition temperature of 81.5°C has been given for the former ester and 75.5°C for the latter. The isotropic liquid was formed at 103.5 and at 82°C and above, respectively (10,11).

EXPERIMENTAL

Campesteryl, sitosteryl and stigmasteryl palmitates were obtained from Applied Science Laboratories, State College, Pennsylvania, in purity reportedly higher than 95%. This was also confirmed by estimation from the shape of DSC traces (12). Purity determined by this method is thought to be the lowest possible value (12). Because of the small quantities available of these dear materials, the major effort to further purify was unwarranted.

Lophenyl and 7-dehydrocholesteryl palmitates were obtained from Institute of Medical Education and Research and Department of Biochemistry, St. Louis University School of Medicine, St. Louis, Missouri. Their isolation and purification are described elsewhere (10,11). The esters, in two to four milligram portions, were precisely weighed and sealed in aluminum planchets followed by analysis for temperatures and heats of transition on a Differential Scanning Calorimeter, Model DSC-2, Perkin-Elmer Corporation. The DSC calibration procedure and the method for evaluation of results have been previously described (13). Results of these measurements are given in Table 1. Sample purities are > 95%.

DISCUSSION

Campesteryl, Sitosteryl and Stigmasteryl Palmitates

Presence of either a methyl or ethyl group at the C-24 position results in the same 13°C increase in the crystalline melting temperature as compared with the reference compound, cholesteryl palmitate. Introduction of a double bond at C-22 in stigmasteryl palmitate leads to an additional 11°C increase in the crystalline melting temperature.

These structural changes in the end chain also affect mesophase formation. Although cholesteryl palmitate exhibits both smectic and cholesteric mesophases, the only mesophase observed for any of the three related sterols is a monotropic smectic mesophase for campesteryl palmitate.

It could be assumed (14) that an alkyl substituent might decrease the molecular interactions due to the added breadth of the molecule and thus increase molecular distances above the level required for cholesteric mesophase formation. Our results are consistent with this hypothesis. The methyl group, however, allows formations of a monotropic smectic mesophase. On the other hand, the more bulky ethyl group does reduce the interactions below the level required for the formation of a smectic mesophase. Previous investigations (5-7, 15) of sitosteryl palmitate did not reveal mesophases either.

The situation does not seem to be so simple, however, as would appear considering only the palmitate esters. It is reported (6) that on cooling, including rapid chilling, of sitosteryl laurate and myristate (though sometimes the melt crystallized directly), a monotropic mesophase resembling the smectic state was detected. Also sitosteryl octanoate was found to form a monotropic smectic mesophase as well as n-heptyl, n-octyl and n-nonyl carbonates (14). It is, therefore, clear that the sitosterol structure facilitates formation of a smectic mesophase.

Considering only fatty acid esters and their mesophase behavior, it is well known that formation of smectic mesophases is dependent also on chain length. With increasing length, the tendency decreases. Mesophases tend to form under certain conditions of high supercooling. Otherwise crystallization from the melt takes place. Moreover, it is not clear how far such mesophase formation is due to possible small amounts of impurities. The lowest temperature of crystallization for sitosteryl palmitate achieved in our experiments was 64°C yet no indication of mesophase formation was observed.

Stigmasteryl esters are a similar case. Investigations of homologues series of stigmasterol fatty acid esters (8), carbonates (16) and thiocarbonates (17) show that most derivatives exhibit monotropic smectic mesophases. A cholesteric mesophase was never observed. For stigmasteryl palmitate, four previous studies did not reveal formation of a mesophase (6,7,9,15). Maidachenko and Chistyakov (8) reported a monotropic isotropic liquid-smectic transition at 61°C, 41°C below the melting point. Attempts were therefore made in this study to achieve the highest possible supercooling. The lowest temperature of crystallization achieved in our experiments and at the highest cooling rates was only 67°C, about 6° above the transition listed by prior authors. No evidence for mesophase formation was found. It should be pointed out that stigmasteryl palmitate was the highest member in the fatty acid ester series for which Maidachenko and Chistyakov detected a smectic mesophase.

Lophenyl and 7-Dehydrocholesteryl Palmitates

Two enantiotropic mesophase transitions have been observed for each ester. Typical DSC traces showing a good resolution of these mesophase transitions are given in Figure 2. The highest temperature mesophase is stable only within a narrow temperature range. It amounts to 1.7^0 for lophenyl palmitate and 2.1^0 for 7-dehydrocholesteryl palmitate. Previous investigations of mesomorphic behavior for several sterols and triterpenes possessing a Δ^7 double bond did not show any formation of a cholesteric mesophase. It was assumed that Δ^7 double bond may inhibit mesophase formation (11). Calorimetric results, however, suggest that moiety does not prevent formation of a cholesteric mesophase but restricts its stability to only a narrow temperature interval.

The entropy changes for all mesophase transitions are larger than the corresponding values found with cholesteryl palmitate [ΔS for smectic-cholesteric and cholesteric-isotropic liquid transitions; 1.04-1.32 and 0.78-0.89 cal/mol/^{0}K respectively (1,18-21)]. The entropy changes associated with both mesophase transitions in lophenyl and 7-dehydrocholesteryl esters amount to approximately 6 and 7.5% from the total solid-isotropic liquid transitional entropy. On the other hand, the entropy change for the single isotropic liquid-smectic mesophase transition in campesteryl ester represents 9.3% of the total entropy change. It demonstrates a more ordered smectic mesophase formed at high supercoolings (18^0 below the solid-isotropic transition).

ACKNOWLEDGEMENT

The authors acknowledge the support of the Materials Research Laboratory at the University of Massachusetts. They also thank Dr. Harold J. Nicholas, St. Louis University School of Medicine, for providing samples of palmitate esters of lophenol and 7-dehydrocholesterol.

REFERENCES

1. G. W. Gray, J. Chem. Soc. 3733 (1956)
2. G. W. Gray, Molecular Structure and Properties of Liquid Crystals, Academic Press, New York, 1962.
3. R. S. Porter, E. M. Barrall, II and J. F. Johnson, Accounts Chem. Res. 2, 53 (1969).
4. G. J. Davis, R. S. Porter and E. M. Barrall, II, Mol. Cryst. and Liq. Cryst. 11, 319 (1970).
5. F. E. King, T. J. King, K. G. Neill and L. Jurd, J. Chem. Soc. 1192 (1953).
6. A. Kuksis and J. M. R. Beveridge, J. Org. Chem. 25, 1209 (1960).
7. F. F. Knapp and H. J. Nicholas, Liquid Crystals and Ordered Fluids, J. F. Johnson and R. S. Porter, Eds., Plenum Press, New York, 1970, p. 147.
8. G. G. Maidachenko and I. G. Chistyakov, Zh. Obsch. Khim. 37, 1730 (1967).
9. A. Heiduschka and H. W. Gloth, Arch. Pharm. 253, 415 (1915).
10. A. M. Atallah and H. J. Nicholas, Mol. Cryst. and Liq. Cryst. 18, 321 (1972).

11. A. M. Atallah and H. J. Nicholas, Mol. Cryst. and Liq. Cryst. $\underline{19}$, 217 (1973).
12. G. J. Davis and R. S. Porter, J. Therm. Anal. $\underline{1}$, 449 (1969).
13. R. J. Krzewki, R. S. Porter, A. M. Atallah and H. J. Nicholas, accepted, Mol. Cryst. and Liq. Cryst.
14. J. L. W. Pohlmann, Mol. Cryst. and Liq. Cryst. $\underline{8}$, 417 (1969).
15. C. W. Griffen, unpublished results.
16. J. L. W. Pohlmann, Mol. Cryst. $\underline{2}$, 15 (1966).
17. W. Elser, Mol. Cryst. and Liq. Cryst. $\underline{8}$, 219 (1969).
18. G. J. Davis, R. S. Porter and E. M. Barrall, II, Mol. Cryst. and Liq. Cryst. $\underline{10}$, 1 (1970).
19. R. D. Ennulat, Mol. Cryst. and Liq. Cryst. $\underline{8}$, 247 (1969).
20. R. S. Porter, E. M. Barrall, II and J. F. Johnson, J. Chem. Phys. $\underline{45}$, 1452 (1966).
21. P. J. Sell and A. W. Neumann, Z. physik. Chem. (N.F.) $\underline{65}$, 13 (1969).

TABLE I

TRANSITION TEMPERATURES AND CALORIMETRY

Palmitate Ester Transition	Temp °C	ΔH cal/g	ΔH kcal/mol/°K	ΔS cal/mol	ΔH Total Solid-Isotropic kcal/mol	ΔS Total Solid-Isotropic cal/mol/°K
Campesteryl						
Solid-Isotropic	91.2	24.6	15.7	43.1	15.7	43.1
Smectic-Isotropic	72.4	2.17	1.39	4.01		
Sitosteryl						
Solid-Isotropic	91.2	24.2	15.8	43.3	15.8	43.3
Stigmasteryl						
Solid-Isotropic	102.6	27.6	18.0	47.8	18.0	47.8
7-Dehydrocholesteryl						
Solid-Smectic	80.5	20.6	12.8	36.2	13.8	39.3
Smectic-Cholesteric	97.2	1.21	0.75	2.03		
Cholesteric-Isotropic	99.3	0.63	0.39	1.05		
Lophenyl						
Solid-Smectic	75.5	21.1	13.5	38.6	14.4	41.1
Smectic-Cholesteric	77.2	0.77	0.49	1.40		
Cholesteric-Isotropic	78.9	0.63	0.40	1.14		

The precision of calorimetric values is believed to be ± 2% and better.

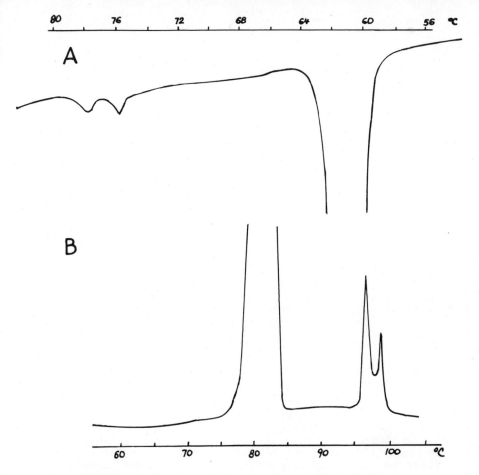

Fig. 1   Structural formulas of (I) cholesterol, (II) campesterol, (III) sitosterol,
(IV) stigmasterol, (V) 7-dehydrocholesterol and (VI) lophenol.

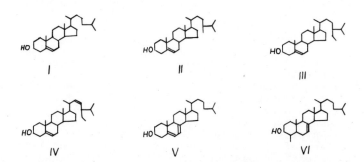

I   II   III

IV   V   VI

Fig. 2   Typical DSC traces for (A) lophenyl palmitate, cooling of isotropic
liquid; (B) 7-dehydrocholesteryl palmitate, heating of solid.

APPLICATION OF LASER FLASH CALORIMETRY FOR HEAT CAPACITY MEASUREMENT ON SAMPLES OF SMALL QUANTITY

Yoichi TAKAHASHI

Department of Nuclear Engineering, University of Tokyo,
Hongo, Bunkyo-ku, Tokyo, 113. Japan

ABSTRACT

The present status and applicability of laser flash method for heat capacity measurement are described. The merits of laser flash method, such as the small quantity of samples necessary for the measurement, are pointed out.

An improved apparatus for the measurement of heat capacity of samples of 0.02-5g is described. In order to improve the accuracy and applicability of the measurement, an "absorbing disk" was emplyed to receive the laser energy impinged. The use of the absorbing disk also enables to measure the heat capacity anomalies which are caused by phase transitions. The precision of the measurement for pellet samples is ±0.5% from 100 to 700K. The experimental results on some nuclear materials such as UP and US are presented. Some preliminary experiments on powder samples encapsulated in an aluminum container are also reported, which show the feasibility of the laser flash technique as a complementary method for thermal analysis.

INTRODUCTION

The laser flash method[1] has been known as a convenient method for measuring thermal diffusivity at high temperatures. It was recognized earlier that one of the advantages of this method was to give heat capacity data simultaneously. In fact, several papers have been published which gave the heat capacity data complementally for the purpose of obtaining thermal conductivity.

It should be emphasized that the laser flash method for heat capacity measurement has the merit of (1) the samll sample size necessary (0.1-5 grams) and (2) the feasibility to do the measurement over a rather wide temperature range by the same apparatus. However, thses advantages have been offset in many of the past studies by the unreliability of the resulting values[2,3] as compared with those obtained by adiabatic calorimetry. The unreliability of the data obtained may be associated with a number of problems encountered with the laser flash method. The main difficulties appear to lie in obtaining:

(1) Accurate measurement of the energy of the laser beam.
(2) Adjustment of the differences in the reflectivity of the

surface of the sample.
(3) Accurate measurement of the temperature rise of the sample
    at high temperatures.

    We have endeavored for several years to solve these
problems and reported methods of measurement to obtain reliable
heat capacity data.[4,5,6] At the first stage of our investigation, the precision of the measurements was not fully satisfactory, and the experimental results on Al_2O_3[4] at temperatures ranging from 300 to 1000K agreed with the NBS standard
data within ±1.5%. The heat capacity was also measured by
this method on UN[5] and US[7] with the same precision.

    Recently, we have newly designed and constructed an
improved laser flash apparatus, by which the heat capacity of
a very small amount (less than 100mg) of powdery sample in an
aluminum capsule can be measured. The precision of heat
capacity determination for pellet samples such as UP and US
was also considerably increased to ±0.5% from 80 to 700K[6].
This paper deals with the capability of the laser flash method
for the measurement of heat capacity and enthalpy increment
of the samples, the mass of which is limited to a rather small
amount.

EXPERIMENTAL TECHNIQUE

Apparatus and its principle

    As the conventional experimental technique of thermal
property measurements by laser flash method was described in
detail in our previous paper[4], the essentials of our improved
technique will be given here. The improved apparatus for
the measurement is shown in Figs. 1 and 2. Samples to be
measured are in the form of a small disk pellet, 8-12mm in
diameter and 0.5-5mm thick. In the case of powder sample, it
is encapsulated in an aluminum container of the same size as
above, which has been used for DSC measurement. The mass of
the sample necessary for the measurement is 0.05-5 grams. On
the front surface of the sample, a thin plate of "absorbing
disk" is attached, the rloe of which is described in detail
in the next section.

    The surface of the absorbing disk is heated with a laser
beam pulse, the duration time of which is about 1 msec, and
the temperature rise of the sample is detected with a copper-
constantan or a Chromel-constantan thermocouple, attached to
the back surface of the sample with silver paste. The electromotive force of the thermocouple is measured by a high-
precision digital micro-voltmeter with a precision of ±0.1μV,
and is recorded by a digital printer twice a second.

    The maximum temperature rise of the sample, ΔT, is
determined from the graphical plotting of the emf, as shown
in Figs. 3 and 4. In the measurement of ΔT, it is essential
that the heat loss from the sample be minimized, and also
an amplifier system of good performance should be used. In
order to minimize the heat loss through conduction along the
thermocouple, couple wires of 0.05mm diameter were used. The
sample was held in a holder made of quartz, designed to
minimize the contact with the sample. The experimental error
in the measurement of ΔT is estimated to be within ±0.2% for

9

the pellets. For the sample of powder form, encapsulated in
an aluminum container, a longer time duration is required to
determine ΔT by extrapolation, as the thermal conduction
within the sample powder is rather slow. In this case, a plot
of log ΔT vs. time elapsed, as shown in Fig. 4, gives better
extrapolation.

In order to determine the energy of the laser flash in
each run of the measurement, a certain portion (several per
cent) of the flashed laser beam is reflected by a glass
plate on a Si-photoelectric cell. The amount of energy
absorbed by the Si-photoelectric cell is indicated by an
electronic integrator. The radiated energy is assumed to be
in linear relation to the energy absorbed by the photo-
electric cell. This was confirmed experimentally, and the
precision of the energy determination is believed to be within
±0.5% or less.

In the presnet method, the molar heat capacity of the
sample, Cp (cal $\cdot$ mol$^{-1} \cdot$ K^{-1}) is given in the following
equation,

$$Cp = \frac{1}{m} \left(\frac{E}{\Delta T} - C \right) \tag{1},$$

where E is the total energy of the laser flash absorbed by
the absorbing disk and the sample (cal), m the quantity of
the sample (mol), and C the heat capacity of the absorbing
disk (cal $\cdot$ K^{-1}). The heat capacities of the aluminum
container, if it is used, of silver paste or silicone grease
and of the thermocouple should be included in the term C.
The term C represented less than 5% of the total for pellet
samples such as UP and US, whereas it becomes up to 40% in
the case of encapsulated samples.

Use of absorbing disk

One of the main improvement of the present work is the
employment of a thin "absorbing disk". The absorbing disk,
being made of glassy carbon of about 5omg and 12mm in diameter
and 0.2mm thick, is attached on the sample surface with very
small amount (1mg) of silver paste or silicone grease, and it
plays the role of absorbing the impinged laser energy.

The use of the absorbing disk makes several advantages in
precise determination of the heat capacity, which are as
follows:

(1) Its size is designed to have a diameter slightly
larger than that of the laser beam, so as to receive all
of the laser beam, which makes easier to get reproducible
results.

(2) The differences in the absorption efficiency of the
laser energy which are caused by the variant emissivity
(reflectivity) of the samples can be avoided by the use of
the absorbing disk, which is made of a definite material, i.e.
glassy carbon. This makes it possible to determine the
absolute heat capacity of the samples.

(3) It has been pointed out that for the samples having
phase transitions, the heat capacity determination by laser
flash method is difficult at the temperature region of the
transitions, because the front surface of the sample is heated
up markedly at the moment of the laser irradiation, from the

10

temperature below transition to that above transition. Then, it is cooled down very quickly, again through transition from high to low temperature, by heat-transfer to inside of the sample. By the use of the absorbing disk, this intense increase of temperature occurs only at the surface of glassy carbon, thus the sudden up and down temperature cycle through the transition can be avoided. Furthermore, because of the improvement of the precision in the measurement of temperature, ΔT can be chosen as small as 0.5K or less at the transition region. The improved method has been proved experimentally to be able to determine the heat capacity anomaly fairly precisely. Examples will be shown later on the magnetic transitions of UP and US.

EXPERIMENTAL RESULTS

Pellet samples

(1) Alumina (Al_2O_3) as the standard material[6]. The sample used was sintered alumina (Wesgo alumina: Al 995), in the form of a disk 9mm in diameter and 0.9mm thick. The deviations of the obtained heat capacity values from those of the NBS standard data[8] are plotted in Fig. 5 from 80 to 700K. In this case, the absolute determination of E in the eq. (1) was conducted by using the literature values of alumina[8] and glassy carbon[9]. The good fit of the experimental data with the standard data without any correction from 200 to 700K proves that this determination of E was quite reasonable.

(2) Uranium monophosphide (UP) and Uranium Monosulfide (US). The heat capacity of UP has not yet been well established at high temperatures, although its determination by the laser flash method was reported by Moser and Kruger[1] in their pioneering work. Its low temperature heat capacity was reported by Counsell, et al.[10] The heat capacity of US from 1.5 to 350K was reported by Westrum et al.[11]

The samples of UP and US used had compositions of $UP_{1.03}$ and $US_{1.00}$ and weighed 3.5-3.8g. The results of the heat capacity determination are shown in Figs. 6 and 7. A sharp anomaly in the heat capacity of UP at 121K, caused by the autiferromagnetic transition was observed. For US, a more diffuse anomaly from 100 to 220K with a peak at 180K was found. In both cases, the obtained results agreed very well with those obtained by adiabatic calorimetry[10,11]. Considering the differences of the amounts of the samples used (15-35g for adiabatic calorimetry), this agreement in the region of heat capacity anomalies is striking. The measurements on UP were extended up to 1000K[12]. The precision of the measurement became worse as the temperature was raised, $\pm 1.5\%$ at 1000K.

Powder samples

The experiments on powdery samples are still in progress, and Fig. 8 shows the results of some preliminary experiments. The sample of powdery Al_2O_3 was encapsulated in a small aluminum container.

The mass of the sample was 20-50mg, and that of the container was about 15mg. The container was sealed off under the atmospheric pressure, and the air in the container facilitated the thermal equlibration of the sample. The deviations of the present results from NBS standard data[8] are about 1.5%. Considering the small mass of the sample used, we believe this results are fairly satisfactory.

ACKNOWLEDGEMENT

The author expresses his gratitude to Dr. M. Murabayashi, Messrs. M. Kamimoto and H. Yokokawa for their collaborations, and his deep appreciation is due to prof. Mukaibo, Univ. of Tokyo, for his helpful discussions. The author also thanks Toray Science Foundation for the partial financial support of this research endeavor.

REFERENCES

1. J. B. Moser and O. L. Kruger, J. Appl. Phys., 38 (1967) 3215.
2. J. B. Moser and O. L. Kruger, J. Nucl. Materials, 17 (1965) 153.
3. S. Nasu and T. Kikuchi, J. Nucl. Sci. Technol., 5 (1968) 54.
4. M. Murabayashi, Y. Takahashi and T. Mukaibo, J. Nucl. Sci, Technol., 7 (1970) 312.
5. Y. Takahashi, M. Murabayashi, Y. Akimoto and T. Mukaibo, J. Nucl. Materials, 38 (1971) 303.
6. Y. Takahashi, J. Nucl. Materials, 51 (1974) 17.
7. M. Murabayashi, Y. Takahashi and T. Mukaibo, J. Nucl. Materials, 40 (1971) 353.
8. D. C. Ginnings and G. T. Furukawa, J. Am. Chem. Soc., 75 (1953), 522.
9. Y. Takahashi and E. F. Westrum, Jr., J. Chem. Thermodynamics, 2 (1970) 847.
10. J. F. Counsell, R. M. Dell, A. R. Junkison and J. F. Martin, Trans. Faraday Soc., 63 (1967) 72.
11. E. F. Westrum, Jr., R. R. Walters, H. E. Flotow and D. W. Osborne, J. Chem. Phys., 48 (1968) 155.
12. H. Yokokawa, Y. Takahashi and T. Mukaibo, presented at the IAEA Symposium on "Thermodynamics of Nuclear Materials" at Vienna, October 1974.

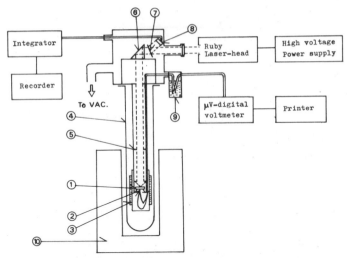

Fig. 1. Schematic diagram of appratus of laser flash calorimetry: 1 Sample, 2 Thermocouple, 3 Internal heater, 4 Quartz container, 5 Adjusting slit, 6 Prîsm, 7 Reflecting glass, 8 Si-photo electric cell, 9 Ice bath, 10 Outer heater or liq. N_2 bath.

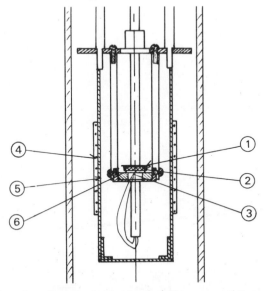

Fig. 2. Sample and its holding assembly: 1 absorbing disk, 2 Sample (or sample capsule), 3 Thermocouple wire, 4 Heater wire, 5 Copper thermal shield, 6 Quartz pin.

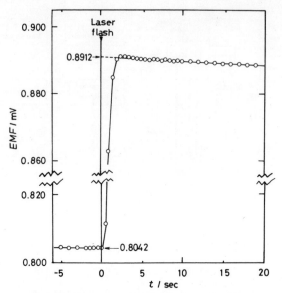

Fig. 3. Determination of maximum temperature rise
(I): Sample UP pellet (3.5501g),
T_1=293.573K, T_2=295.724K, ΔT=2.151K.

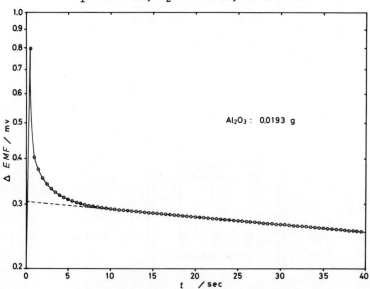

Fig. 4. Determination of Maximum temperature
rise (II): Sample Al_2O_3 powder (0.0193g)
in Al capsule (0.0171g), T_1=290.223K,
T_2=295.299K, ΔT=5.076K.

14

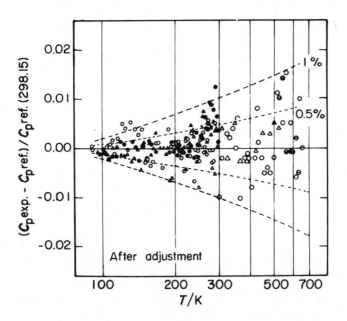

Fig. 5. Deviation of the heat capacity of Al_2 (pellet) (Cp_{exp}) from NBS standard data (Cp_{ref}).

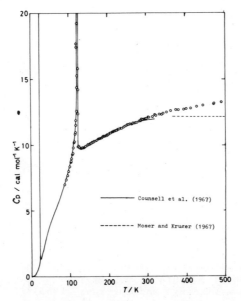

Fig. 6. Heat capacity of $UP_{1.03}$ (pellet; 3.5501g)

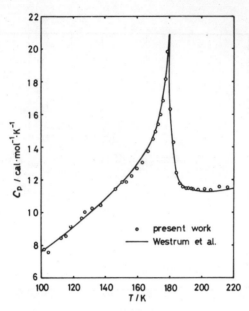

Fig. 7.   Heat capacity anomaly of $US_{1.00}$ (pellet;
3.7671g). ———: Westrum et al.
(1968).[11]

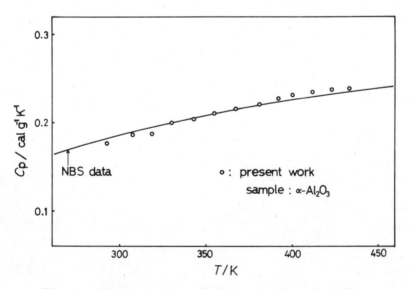

Fig. 8.   Heat capacity of Al_2O_3 (powder; 0.0193g
in Al capsule of 0.0171g).
———Ginnings and Furukawa (1953)[8].

SOME THERMOANALYTICAL APPLICATIONS TO THE CHEMISTRY OF COORDINATION COMPOUNDS

Patrick K. Gallagher

Bell Laboratories
Murray Hill, New Jersey   07974

INTRODUCTION

Defining thermal analysis and limiting its scope to a practical workable range is difficult. Add to this the even broader term, coordination chemistry, and you obtain an immense, complex, and somewhat ill defined topic. In order to give some attention to detail and the finer points, I have chosen to limit my discussion of the field to three general classes of compounds (1) the classical example coordination compounds, ammines, (2) cyano-complexes, and (3) oxalates. These will provide ample opportunities to illustrate the value of thermo-analytical techniques to the investigation of such important and interrelated subjects as metal to ligand bonding, electron transfer processes, the stoichiometry and mechanisms of thermal decompositions, and the applicability of such studies to inorganic synthesis.

AMMINE COMPLEXES

Few things in inorganic chemistry have been studied as extensively as Werner complexes such as $[Co(NH_3)_6]^{3+}$. Although spectroscopy and magnetic properties have been the leader in elucidating structure and bonding in the solid state, thermo-analytical techniques have made contributions in recent years.

One area of interest in this general field is whether the water molecules are part of the inner coordination sphere, e.g. $[Co(NH_3)_5(H_2O)]^{3+}$ or present as the more common water of hydra-tion. Not only is this of interest in determining the nature of the initial complex but is also important in following the course of a decomposition. As an example, Nickolaev and coworkers[1] have investigated the thermal decomposition of a number of EDTA complexes, of the type $M_2EDTA \cdot 9H_2O$, and distin-guished between those H_2O which are bound to the metal and those which are merely waters of crystallization. $Mg_2EDTA \cdot 9H_2O$ loses its H_2O simultaneously with a peak in the differential thermogravimetric (DTG) curve at ∼170°C. However, $MgPbEDTA \cdot 9H_2O$ loses $6H_2O$ at a similar temperature but the last $3H_2O$ give a peak in the DTG at 260°C implying that they are more strongly held. Furthermore, the activation energy (ΔH^*) is about 15 Kcal/mole while ΔH^* for the removal of the last

$3H_2O$ from $M_gPbEDTA \cdot 9H_2$ is about 30 Kcal/mole. They had previously deduced that the ΔH^* for the loss of water of crystallization is 15-20 Kcal/m and for H_2O which is coordinately bounded it is more like 32-38 Kcal/m.

Wendlandt and Smith[2] followed the course of dehydration of triaquatriaminecobalt(III) halides. The $3H_2O$ were initially coordinated to the Co and the halide ions were external to the first coordination sphere. However, as the dehydration progressed the departing H_2O were replaced by the halide ions in a stepwise fashion indicated in Eq. (1).

$$[Co(NH_3)_3(H_2O)_3]X_3 \quad \xrightarrow{60°-100°} \quad [Co(NH_3)_3(H_2O)_2X]X_2 + H_2O \uparrow$$

$$\xrightarrow[200°]{100°-} [Co(NH_3)_3X_3] + 2H_2O \uparrow \tag{1}$$

Temperatures of deaquation are quite low here for coordinated water but there is a compensation in energy involved due to the entry of the halide ion into the first coordination sphere. Unfortunately, Wendlandt and Smith did not determine the ΔH^* for either step in the deaquation.

Moving on to ligands other than H_2O. Liptay, et al.,[3] did a nice piece of work on the thermal decomposition of some transition metal pyridine-thiocyanate complexes. Table I gives some selected values for the temperatures at the peak of DTG curves for the reactions:

$$M(py)_4(SCN)_2 \rightarrow M(py)_2(SCN)_2 + 2py \uparrow \tag{2}$$

and

$$M(py)_2(SCN)_2 \rightarrow M(SCN)_2 + 2py \uparrow \tag{3}$$

The thermal stability of these compounds is in direct relation to the strength of their formation equilibrium quotients. Since the decomposition occurs by breaking the metal to pyridine bond it is logical that the stability is based upon the strength of this metal-nitrogen link.

Attack in some chelates, however, is not directly at the metal-ligand bond but frequently at a point within the chelate entity. Under these circumstances it is the influence of the metal ion by induction upon the appropriate bond. As an example the decomposition temperatures of some transition metal chelates are given in Table II along with their formation equilibrium quotients. The opposite effect is observed. When the metal ligand bonding is strong this weakens the bonding within the chelate by an inductive effect. Since the thermal decomposition originates at this point within the chelate ring, the thermal stability is reduced by strong metal-chelate bonding.

18

The final subject which I wish to cover in this general area of ammine complexes involves the use of thermoanalytical technique to follow the reduction of the metal ion during the course of a thermal decomposition. Again Professor Wendlandt and his coworkers have led the way. As an example consider the series of $Co(NH_3)_6{}^{3+}$ complexes.[4-5] The thermal decomposition of these materials was studied by thermal magnetic and evolved gas analyses (EGA) as well as more conventional TG and DTA. Thermal magnetic analysis is interesting in that it allows one to follow the reduction of Co(III) to Co(II) in the presence of other thermal and weight loss effects. Because Co(III) is d^6 and in a low spin octahedral complex involving d^2sp^3 hybridization, it is diamagnetic while the Co(II), d^7, is paramagnetic. By subjecting the sample to a magnetic attraction it will undergo an apparent weight gain as the Co(III) → Co(II) reaction proceeds. Figure 1 is an illustration of the observed effect. The first compound to undergo reduction is the iodide, curve C. The I^- is relatively easily oxidized to iodine and the reaction is

$$2[Co(NH_3)_6]I_3 \rightarrow 2CoI_2 + I_2 + 12(NH_3)\uparrow \qquad (4)$$

The (NH_3), however, is more readily reduced than the Cl^-, Br^-, or $SO_4{}^{2-}$. These compounds consequently tend to reduce at higher temperatures than the iodide according to the following stoichiometry illustrated for the chloride.

$$6[Co(NH_3)_6]Cl_3 \rightarrow 6CoCl_2 + N_2 + 6NH_4Cl + 28NH_3 \qquad (5)$$

Overall stoichiometry for the decomposition of the nitrate is complicated by the oxidizing nature of $NO_3{}^-$ and the reaction occurs abruptly beginning at a somewhat higher temperature, curve E. General conclusions drawn from these thermomagnetic curves are in excellent agreement with other techniques, i.e., TG, DTA, and EGA.

Tanaka and Manjo[6] did a good follow-up on the reduction of $[Co(NH_3)_6]Cl_3$ by investigating both the IR spectra of samples heated at various temperatures and the polaragraphic properties of solutions made from these samples. Polaragraphic analysis provided a quantitative determination of the relative amounts of Co^{3+} and Co^{2+} at different temperatures. IR spectra, on the other hand, provide a means of following the NH_3-$NH_4{}^+$ reaction. Figure 2 shows some of their results. Bands at 1600, 1320, and 830 cm^{-1} correspond to bound (NH_3). They predominate initially but disappear around 305-340°C. The new band at 1400 cm^{-1} corresponds to $NH_4{}^+$ and appears around 290°C and disappears around 400°. Finally the sharp band for free NH_3 becomes evident at the two highest temperatures. These results correlate well with Eq. (5) where the reduction of Co(III) is accompanied by the formation of $NH_4{}^+$ and free NH_3. The latter band is hidden by the broader one from bound (NH_3). At the higher temperature $NH_4{}^+$ decomposes and the bound (NH_3) is completely decomposed leaving only free NH_3.

Analogous Cu(II) complexes have also been studied by Smith and Wendlandt[7] using the same series of techniques. In this case Cu^{2+} is d^9 and paramagnetic while Cu^+ is d^{10} and diamagnetic. Hence, there is an apparent weight loss associated with the reduction in a magnetic field. Figure 3 shows the results. The relative stability of the compound is the same as was observed for the Co salts. Again I^- is the reducing agent when present and (NH_3) served in the absence of I^-

In the previous examples the same reduced state was common to all the compounds. Tetrammine palladium(II) halides on the other hand yield different reduced states of Pd due to the different temperatures at which they decompose.[8] The iodide salt again decomposes at the lowest temperature, 250°C in He, yielding PdI_2. The bromide persists until 430°C in He and then decomposes to give a mixture of $4PdBr_2$ and Pd. Finally, the chloride decomposes directly to give Pd without any evidence of $PdCl_2$ as an intermediate.

Even a single complex can go by two different paths producing different products based on the heating rate and particle size. Joyner and Verhoek[9] have shown that $[Co(NH_3)_6](N_3)_3$ can go via two paths summarized below:

$$2[Co(NH_3)_6](N_3)_3 \rightarrow 2Co + 11N_2 + 8NH_3 + 6H_2 \qquad (6)$$

$$[Co(NH_3)_6](N_3)_3 \rightarrow CoN + 4N_2 + 6NH_3 \qquad (7)$$

The path indicated by Eq. (6) is explosive and is observed on occasions where local heating can be present. The other pathway is slower and controlled. It results from more gradual heating of larger particles such as crushed crystals.

CYANO COMPLEXES

Having discussed various aspects of bonding using ammine complexes as examples I would now like to discuss ways of determining the stoichiometry of a reaction and its intermediate phases. In keeping with the coordination chemistry theme, I would like to use the highly complex thermal decomposition of $EuFe(CN)_6 \cdot 5H_2O$ and $NH_4EuFe(CN)_6 \cdot 4H_2O$, as examples. These are particularly good examples because a large variety of techniques have been used to help elucidate the course of the decomposition.

These and related compounds are of interest for the preparation of stoichiometric rare earth orthoferrites and the analogous Co compounds.[10] The latter compounds are of current interest as potential catalysts for automobile emission control.[11] Desired oxides can be obtained with essentially perfect stoichiometry from the decomposition of the complex cyanide in O_2 or air.[10,12] However, when these complex cyanides are decomposed in vacuum or an inert atmosphere their reactions are complicated and require a battery of weapons from the scientific arsenal in order to understand the process.[12,13]

The Eu compounds were selected for a detailed investigation, prior to surveying the entire rare earth series, for several reasons. Europium is the most readily reduced of the trivalent rare earths and consequently the chemistry associated with the decomposition should prove most interesting. In addition, the Mössbauer Effect can be readily utilized to investigate the rare earth ion in this case as well as the iron ion. This technique has proved useful in investigations of this type, particularly to follow changes in oxidation state.

As stated previously, decomposition of these compounds in O_2 leads to the formation of $EuFeO_3$ in a relatively straightforward manner. Figure 4 shows two weight losses. The first loss corresponds to most of the water of hydration. The compounds ignite around 300°C and burn rapidly to form finely divided intimately mixed Eu_2O_3 and Fe_2O_3. A small particle size can be inferred both from the poor quality of the X-ray patterns in the 300-500°C region (Fig. 5) and the Mössbauer spectra (Fig. 6) which indicate that the Fe_2O_3 is so poorly ordered that it is in a superparamagnetic state. The characteristic six-line pattern of Fe_2O_3 which indicates its antiferromagnetic alingment is absent and instead a doublet is present having an isomer shift associated with Fe^{3+}. This is typical behavior of finely divided or amorphous Fe_2O_3 from a variety of sources.[14,15] Similar Mössbauer spectra are observed starting with $NH_4EuFe(CN)_6 \cdot 4H_2O$ except that the room temperature spectrum is that of the divalent compound, i.e., IS = - 0.30 mm/sec with no quadrupole splitting.

In the region of 600°-700°C, or somewhat earlier starting with the Fe(II) compound, the oxides react to form the orthoferrite. Both the X-ray pattern and the Mössbauer spectrum of this compound are evident. There is also an exothermic peak in the DTA pattern associated with this reaction. The DTA peaks are at somewhat higher temperature due to the rapid scanning rate, 10°C/min, employed for this measurement.

Changes in the surface areas of the samples calcined in air are consistent with this interpretation. The peak in surface area or minimum particle size occurs after the major decomposition around 300°C. The oxides which are formed then tend to anneal, sinter, and react to reduce their surface area. Around 600° to 700° there is a slight change in slope caused by the reaction to form $LaFeO_3$. It would appear that orthoferrites having particles sizes of 0.1 μ or greater can be prepared depending upon the choice of calcined temperature. Such finely divided material is a value in the catalytic applitions or where further reactivity such as for sintering is important.

In contrast to this simplicity, Fig. 7 shows the results of thermal analysis in an inert atmosphere. The complexity of these curves compared to those in Fig. 4 is obvious. Unfortunately, as is frequently the case X-ray diffraction analysis is of little help in determining the different stages of the decomposition at intermediate temperatures because the products are either microcrystalline or virtually amorphous due to the

21

turmoil and major readjustments brought about by the decomposition. EGA (Figs. 8 and 9),[13] IR spectra (Fig. 10),[13] and Mössbauer spectra (Figs. 11 and 12)[12] allow one to describe the reaction reasonably well. The first major step in the decomposition of either compound is hydrolytic in nature and also involves a reduction of the Fe(III) for the $Fe(CN)_6^{3-}$ complex. Equations (8) and (9)

$$2EuFe(CN)_6 \rightarrow 2EuOOH + 2Fe(CN)_2 + (CN)_2 + 6H_2O + 6HCN \quad (8)$$

$$NH_4EuFe(CN)_6 \rightarrow EuOOH + Fe(CN)_2 + NH_3 + 2H_2O + 4HCN \quad (9)$$

describe this first step. Reduction of Fe(III) is clearly obvious from the Mössbauer spectrum at $300^\circ C$ in Fig. 11. The quadrupole split doublet associated with Fe(III) has disappeared and instead there is a single line with an isomer shift corresponding to a divalent cyanide complex similar to the room temperature spectrum of $NH_4EuFe(CN)_6 \cdot 4H_2O$. IR spectra of the C≡N bands in Fig. 10 also substantiate the reduction going from the room temperature to the sample heated at 300°C. This reduction is accomplished by the evolution of $(CN)_2$ which can be seen in the EGA, Fig. 8. Since there is no reduction involved in the decomposition of the $NH_4LaFe(CN)_6$, there is no analogous peak in Fig. 9.

The hydrolytic nature of the decomposition can be discerned from the large evolution of HCN in Figs. 8 and 9 for both compounds. Resulting broad OH bands are also obvious in the IR spectra, Fig. 10. Decomposition of the NH_4^+ is also evident from the EGA curve corresponding to NH_3 in Fig. 9.

The EuOOH which is formed by the hydrolytic decomposition loses water over a wide temperature range as evidenced by the persistence of the bands associated with OH in Fig. 10. This is consistent with the work of Rau and Glover.[16] However, in a vacuum and in the presence of carbon or carbides the Eu^{3+} becomes partially reduced at elevated temperatures. This is evident from the Mössbauer spectra in Fig. 12. The absorption around -13 mm/sec is due to Eu^{2+}. It increases with increasing temperature.

The divalent iron cyanide, which I have somewhat arbitrarily designated as $Fe(CN)_2$ as opposed to say $Fe_2Fe(CN)_6$, decomposes above 600°C to form Fe_3C. This is primarily based upon Mössbauer spectra in Fig. 11 and is also consistent with EGA and IR spectra. An isomer shift of 0 to +0.06 mm/sec and a magnetic hyperfine splitting of 207-210 kOe are in excellent agreement with the reported values for Fe_3C.[17] The $Fe(CN)_2$ to Fe_3C reaction involves the evolution of N_2. Subsequent formation of Fe from Fe_3C may be a simple dissociation liberating C or may be a reaction with the Eu_2O_3 to form EuO or Eu_3O_4 and CO. Both CO and N_2 would give rise to the large mass 28 peak observed in the EGA, Figs. 8 and 9.

It should be evident from the preceding discussion of the decomposition of these compounds that it is frequently necessary to combine a variety of techniques in order to obtain even such

22

a qualitative picture as this of the course of a complex decomposition.

OXALATES

Besides providing further interesting examples of the application of thermoanalytical techniques to the study of bonding and mechanisms of decompositions, studies of oxalate decompositions also provide a good opportunity to look at some practical aspects as well. Oxalates are particularly useful as synthetic intermediates in the potential production of a variety of both simple and complex oxides. Their complex ion equilibria and solubility offer many advantages in the preparation of a homogeneous precursor compound which has the desired stoichiometry or ratio of cations. Coupling this with their ready thermal decomposition to yield finely divided highly reactive oxides, makes them attractive aids in some synthesis. As examples, the precipitation of such compounds as $BaTiO(C_2O_4)_2$, $BaSn(C_2O_4)_2$, and $Sr_3Fe_2(C_2O_4)_6$ offer paths for the precise control of stoichiometry for the preparation of $BaTiO_3$,[18] $BaSnO_3$,[19] and $Sr_3Fe_2O_7$.[20] Similarly, precipitation of a solid solution of divalent transition metal oxalates provides mixing on an atomic scale vastly preferable to what might be achieved by conventional mixing of oxides or carbonates in the synthesis of ferrites.[21,22] Even when solubility does not allow precipitation, such techniques as freeze drying or spray drying of appropriate solutions can yield an oxalate with intimately mixed components for subsequent processing, e.g., $LiNbO_3$[23] or Fe_2O_3.[24] Control of time temperature, and atmosphere during the calcination of oxalates affords the opportunity to control the particle size and to some extent the reactivity of the final oxide.

Before considering an example of the synthetic usefulness in greater detail, however, I would like to discuss one of the particularly interesting but unfortunately complicating aspects of the decomposition of oxalates. This is the uncertain extent of the disproportionation of any CO evolved. Normally CO is a relatively stable gas which from an environmental point of view is an all too stable contaminant. In fact, however, it is thermodynamically highly unstable below 500°C and persists only because of kinetic factors. Finely divided oxides, especially of transition metals, are generally catalytically active. Consequently, if the evolved gas is allowed to remain in contact with the solid products, the disproportionation of CO will proceed to an unfortunately unpredictable extent.

$$2CO \rightleftarrows CO_2 + C \qquad (13)$$

Figure 13 is an indication of the equilibrium of this reaction at atmospheric pressure of the combined compounds and is based on data tabulated by K. K. Kelley.[25] Obviously this reaction renders the interpretation of evolved gas analysis difficult and carbon content frequently darkens the samples and presents a problem in thermogravimetry. If the decomposition is carried out in an oxidizing atmosphere this problem is minimized because the CO is oxidized to CO_2. This is a highly exothermic reaction, however, which completely overwhelms the endothermic decomposition of the oxalate in DTA studies and also can lead to substantial heating of the sample.

An interesting synthetic aspect of the disproportionation of CO is that it may be used to form carbides at relatively low temperatures. TaC for instance can be formed at 1200°C by decomposing the oxalate in a reducing atmosphere.[23] Decomposition under such conditions provides a finely divided highly reactive C from the disproportionation which is intimately mixed with a reduced form of the metal oxide.

Finally I would like to demonstrate the advantages of using oxalates as precursors in synthesis by studying the stoichiometry and kinetics of a particular example in some detail. Barium stannate, $BaSnO_3$, is commonly used as a source for investigations of the Mössbauer Effect in Sn compounds. It has one drawback for use as a standard, however, which is the uncertainty in its precise stoichiometry. The purpose therefore, was to devise a preparative technique for $BaSnO_3$ which would assure the following: an atomic ratio for the Ba to Sn of unity, the proper oxygen content, and high purity.

Precipitation of a unique compound analogous to barium titanyl oxalate[18] offered an excellent approach. Precipitation of such a compound should result in the desired atomic ratio since the excess of either Ba or Sn would remain behind in solution. The technique also yields a precursor compound which could be readily converted at a relatively low temperature to the desired oxide having a high purity. This low temperature conversion is important because it alleviates much of the uncertainty in oxygen content associated with the more conventional preparation from the high temperature reaction of the oxides.

Complex oxalates of the composition indicated in Table III and IV were prepared from Sn^{4+} and Sn^{2+} respectively. These tables also summarize the stoichiometric chain of intermediates in both O_2 and N_2 based upon the summation of the results of TG, DTA, EGA, X-ray diffraction studies and Mössbauer spectroscopy.

The chain of intermediates in Table III outlines the general scheme of the thermal decomposition in O_2. Endotherms below 250°C are due to dehydration and suggest a multistep process. The TG curve seems quite consistent with two equal steps of three water molecules.

Presence of $Sn(CO_3)_2$ is speculative based on the inflection point in the TG curves. Unfortunately, the decomposition products are so amorphous or finely divided at these low temperatures that they give no X-ray diffraction patterns to aid in their characterization. A carbonate is a reasonable and frequent intermediate during the decomposition of an oxalate.

24

A point of major interest is whether the Sn undergoes a reduction during the oxalate decomposition. Mössbauer spectra in Fig. 14 clearly show the absence of Sn^{2+} in the samples calcined in an oxidizing atmosphere. There is significant Sn^{2+} in the sample calcined at 300°C in vacuum. There are also two exothermic peaks in the DTA patterns obtained in the N_2. These must be associated with either an amorphous to crystalline transition or oxidation of Sn^{2+}. The absence of O_2 precludes the oxidation of CO which is responsible for much of the exothermic nature of the DTA pattern determine in O_2. X-rays are of no use in identifying the low temperature intermediate and consequently in specific divalent intermediate in the decomposition in vacuum is unknown.

The last two stages of the decomposition are of particular interest since they involve the desired end product, $BaSnO_3$. The $BaCO_3$ reacts in a stepwise fashion as indicated by the TG, X-ray, and Mössbauer spectra. Conversion of SnO_2 to $BaSnO_3$ is complete around 800°C depending upon the atmosphere. Mössbauer spectra in Fig. 14 show that the slight quadrupole splitting present initially has disappeared at 700°C and a sharp single line pattern results. The additional mole of $BaCO_3$ is no major problem in its use as a Mössbauer source or absorber. However, if it is desired to heat treat it further to improve the crystallinity it would be advisable to leach out the $BaCO_3$ with acid. At higher temperatures the $BaCO_3$ will react with the $BaSnO_3$ to form Ba_2SnO_4 which has an undesirable quadrupole splitting as indicated in Fig. 14d. These high temperature reactions are not particularly evident in the DTA patterns. The peak around 820°C is associated with a phase transition in $BaCO_3$.

The basic scheme of the thermal decomposition of the divalent compound is similarly outlined by the chain of intermediates presented in Table IV. In O_2 the Mössbauer spectra shown in Fig. 15 clearly indicate that Sn^{2+} is virtually completely converted to the tetravalent state by 400°C. The DTA pattern suggests that in O_2 there is a large broad exothermic peak associated with this oxidation. Superimposed upon this are two endothermic peaks attributed to the dehydration and oxalate decomposition. Oxidation of the CO evolved also contributes significantly to the intense broad exotherm. Only the two endotherms are present in N_2.

Because of the stoichiometry of the oxalate, $BaSnO_3$ is the end product uncontaminated with $BaCO_3$. Consequently, it can be given extensive heat treatments to improve its crystallinity, but, it must be remembered that the elevated temperature will induce a higher concentration of oxygen vacancies. The sample heated to 1000°C shows neither Sn^{2+} nor significant line broadening in Fig. 15d.

Decomposition in vacuum gives metallic Sn as an intermediate. This is evident not only from the X-ray pattern but also in the Mössbauer spectra.

It is interesting that, although both decomposition yield mixtures of $BaCO_3$ and SnO_2, there is no evidence for the formation of Ba_2SnO_4 prior to the complete formation of $BaSnO_3$. This is contrary to that observed for physical mixtures[26] and consequently led to kinetic studies of these reactions.[27]

Figure 16 shows the relevant X-ray diffraction patterns. The oxalate compound decomposes to yield $BaCO_3$ and SnO_2 which react cleanly and quickly between 600° and 700°C to form $BaSnO_3$ without the formation of intermediate phases. The $BaCO_3 \cdot SnO_3$ pattern at 1000°C clearly shows the presence of $BaSnO_3$, Ba_2SnO_4, and SnO_2 indicating the presence of Ba_2SnO_4 as an intermediate phase.

The several hundred degrees difference between the two reaction zones arises from the difference in the pre-exponential term and probably reflects the relative differences in particle sizes and contact areas. In the coprecipitated compound the Ba and Sn are mixed on an atomic scale. This, however, cannot be maintained after decomposition, as shown by the presence of the X-ray diffraction patterns of $BaCO_3$ and SnO_2 in Fig. 16a. The breadth of the lines do indicate that the particles are small and no doubt strained with a high concentration of physical defects. The homogeneity of such a mix must also surely surpass that obtained by physical mixing. Consequently, this increased surface area and points of contact give rise to an almost homogeneous-type second-order reaction with a relatively high pre-exponential term. In contrast the physical mixture of the bulk materials is made up of coarser particles which are not nearly as intimately mixed. Their reaction is described by geometrical rate laws and is slower even though the same bonds are broken and formed during the reaction. The reaction also requires a longer diffusion path and thereby creates regions of Ba_2SnO_4.

SUMMARY

Conventional thermoanalytical techniques can play an important role in elucidating the bonding and structure of complexes. Coupled with a variety of other analytical techniques they can help define the stoichiometry, kinetics, and mechanisms of the decompositions of these materials. Understanding these aspects is highly desirable in utilizing such complex coordination compounds as precursors in inorganic synthesis.

REFERENCES

1. A. V. Nikolaev, V. A. Lagvinenko, and L. I. Myachina, Thermal Analysis Vol. II, Edited by R. F. Schwenker, Jr. and P. D. Garn, Academic Press, N.Y. 1969, p. 779.

2. W. W. Wendlandt and J. P. Smith, J. Inorg. Nuclear Chem., 26, 1619 (1964).

3. G. Liptay, E. Papp-Molnár, and K. Burger, J. Inorg. Nuclear Chem., <u>31</u>, 247 (1969).

4. W. W. Wendlandt, J. Inorg. Nuclear Chem., <u>25</u>, 545 (1963).

5. W. W. Wendlandt and J. P. Smith, J. Inorg. Nuclear Chem., <u>25</u>, 1267 (1963).

6. N. Tanaka and M. Manjo, Bull. Chem. Soc., Japan, <u>37</u>, 1330 (1967).

7. J. P. Smith and W. W. Wendlandt, J. Inorg. Nuclear Chem., <u>26</u>, 1157 (1964).

8. W. W. Wendlandt and L. A. Funes, J. Inorg. Nuclear Chem., <u>26</u>, 1879 (1964).

9. T. B. Joyner and F. H. Verhoek, Inorg. Chem., <u>2</u>, 334 (1963).

10. P. K. Gallagher, Mater. Res. Bull., <u>3</u>, 225 (1968).

11. D. W. Johnson, Jr. and P. K. Gallagher, Thermochimica Acta, <u>7</u>, 303 (1973).

12. P. K. Gallagher and F. Schrey, "Thermal Analysis," Vol. 2, R. F. Schwenker and P. D. Garn, Ed., Academic Press, New York, 1969, p. 929.

13. P. K. Gallagher and B. Prescott, Inorg. Chem., <u>9</u>, 2510 (1970).

14. P. K. Gallagher and C. R. Kurkjian, Inorg. Chem., <u>5</u>, 214 (1966).

15. T. Nakamura, T. Shinjo, Y. Endoh, N. Yamamoto, M. Shiga, and V. Nakmura, Phys. Letters, <u>12</u>, 178 (1964).

16. R. C. Rau and W. J. Glover, Jr., J. Am. Ceram. Soc., <u>47</u>, 382 (1964).

17. H. Bernas, I. A. Campbell, and R. Fruchart, J. Phys. Chem. Solids, <u>28</u>, 17 (1967).

18. P. K. Gallagher and J. Thomson, Jr., J. Am. Ceram. Soc., <u>48</u>, 644 (1965).

19. P. K. Gallagher and F. Schrey, Thermal Analysis, Vol. 2. Edited by H. G. Weidemann, Berkhauser Verlag, Basel, Switz, 1972 p. 623.

20. P. K. Gallagher, Inorg. Chem., <u>4</u>, 965 (1965).

21. P. K. Gallagher and F. Schrey, J. Am. Ceram. Soc., <u>47</u>, 434 (1964).

22. P. K. Gallagher, H. M. O'Bryan, F. Schrey, and F. R. Monforte, Am. Ceram. Soc. Bull., 48, 1053 (1969).

23. P. K. Gallagher and F. Schrey, Thermochimica Acta, 1, 465 (1970).

24. P. K. Gallagher, D. W. Johnson, Jr., F. Schrey, and D. J. Nitti, Am. Ceram. Soc. Bull., 52, 842 (1973).

25. K. K. Kelley, "Contribution to the Data on Theoretical Metallurgy: XIII High-Temperature Heat-Content, Heat-Capacity, and Entropy Data for Elements and Inorganic Compounds," Bureau Of Mines Bull. 584, U.S. Government Printing Office, Washington, D.C. 1960.

26. G. Von Wagner and H. Binder, Z. Anorg. Allg. Chem., 297, 334 (1958).

27. P. K. Gallagher and D. W. Johnson, Jr., Thermochimica Acta, 4, 283 (1972).

TABLE I

Peak Temperature in the DTG Curves for the Loss of 2py from the Indicated Complexes[3]

Complex	°C	Complex	°C
$Ni(py)_4(SCN)_2$	190	$Ni(py)_2(SCN)_2$	280
$Co(py)_4(SCN)_2$	180	$Co(py)_2(SCN)_2$	270
$Mn(py)_4(SCN)_2$	130	$Mn(py)_2(SCN)_2$	240
		$Zn(py)_2(SCN)_2$	280
		$Cd(py)_2(SCN)_2$	260

TABLE II

Peak Temperatures in the DTG Curves for the Decomposition of the Indicated Complexes and the Formation Constants in H_2O-Dioxane Media[3]

Complex	°C	Log β_2
Copper (II) salicylaldioxane	170	21.5
Iron (II) salicylaldioxane	190	16.7
Nickel (II) salicylaldioxane	240	14.3
Copper (II) Dimethylglyoxime	160	23.5
Nickel (II) dimethylglyoxime	290	21.7

TABLE III

Summary of Thermogravimetric Results for $Ba_2Sn(C_2O_4)_4 \cdot 6H_2O$

Intermediates	Theo. wt.%	Exp. O_2 wt.%	Deg. C	Exp. wt.%	Deg. C
$Ba_2Sn(C_2O_4)_4 \cdot 3H_2O$	93.7	91	RT-160	91	RT-165
$Ba_2Sn(C_2O_4)_4$	87.4	87	160-240	87	165-250
$2BaCO_3 + Sn(CO_3)_2$	74.2	71	240-320	74	250-325
$2BaCO_3 + SnO_3$	63.9	63	320-440	61	325-450
$BaCO_3 + BaSnO_3$	58.8	58	440-730	57	450-850
Ba_2SnO_4	53.6	53.6	730-950	--	850-1000

TABLE IV

Summary of Thermogravimetric Results for $BaSn(C_2O_4)_2 \cdot 1/2H_2O$

Intermediates	Theo. wt.%	Exp. O_2 wt.%	Deg. C	Exp. N_2 wt.%	Deg. C
$BaSn(C_2O_4)_2$	98.0	98	RT-370	--	-
$BaC_2O_4 + SnO$	81.6	-	---	83	RT-425
$BaCO_3 + SnO$	75.3	-	---	74	425-480
$BaCO_3 + SnO_2$	78.9	76	370-400	--	-
$BaSnO_3$	68.9	68.5	400-95C	68.2	480-950

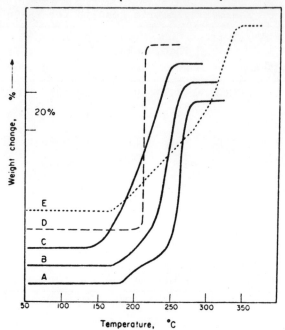

Fig. 1

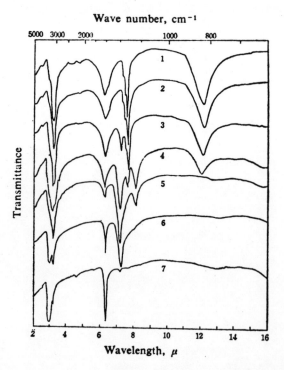

Fig. 2

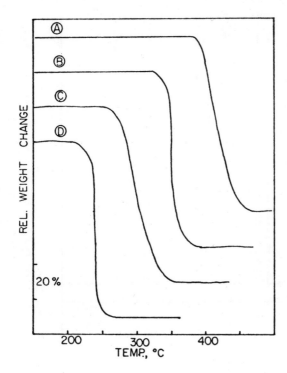

Fig. 3

Fig.1 Thermomagnetic curves of some hexamine cobalt (III)
complexes[5]

    a Cl d NO_3
    b Br e SO_4
    c I

Fig.2 Infrared absorption spectra of $(Co(NH_3)_6)Cl_3$ heated at
various temperatures[6]

    1 Unheated   5 305°C
    2 170°C   6 340
    3 285   7 475
    4 295

Fig.3 Thermomagnetic curves of some hexamine copper (II)
complexes[7]

    a SO_4 c Br
    b Cl d I

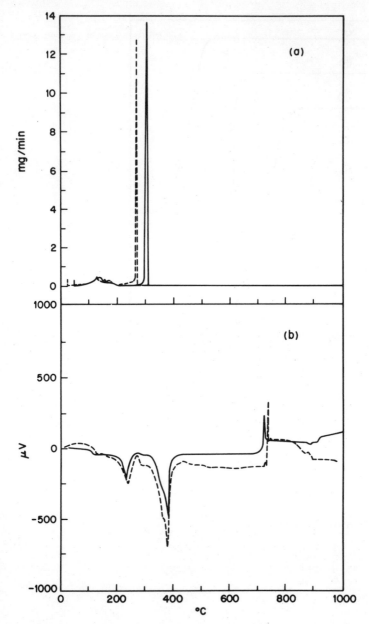

Fig.4 Thermal analysis in oxygen[12]

a	DTG	——	EuFe(CN)$_6$ · 5H$_2$O
b	DTA	- - -	NH$_4$EuFe(CN)$_6$·4H$_2$O

32

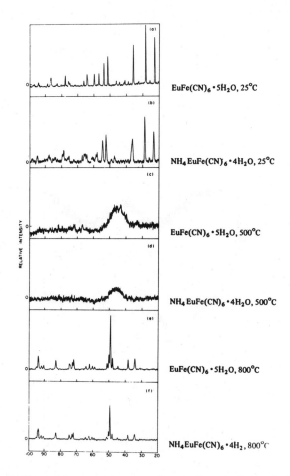

Fig.5 X-ray diffraction patterns of samples calcined in air.

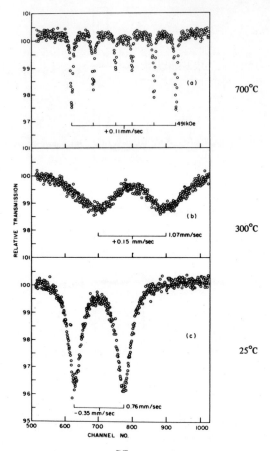

Fig.6 Mössbauer spectra of ^{57}Fe in samples of EuFe$(CN)_6\cdot5H_2O$ calcined in air at various temperatures.[12]

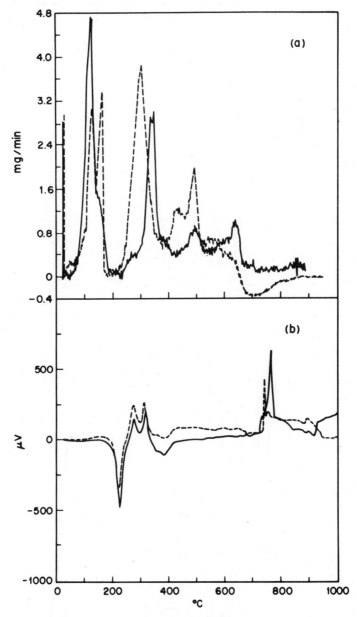

Fig.7 Thermal analysis in hélium

    a   DTG   ——— $EuFe(CN)_6 \cdot 5H_2O$

    b   DTA   - - - - $NH_4EuFe(CN)_6 \cdot 4H_2O$

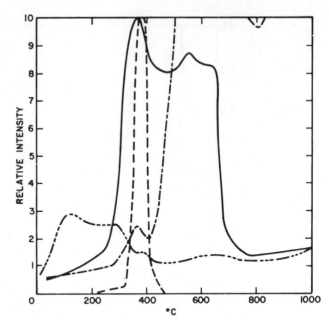

Fig. 8.

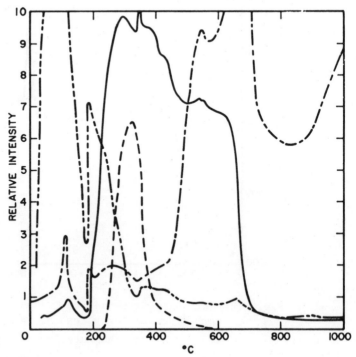

Fig. 9.

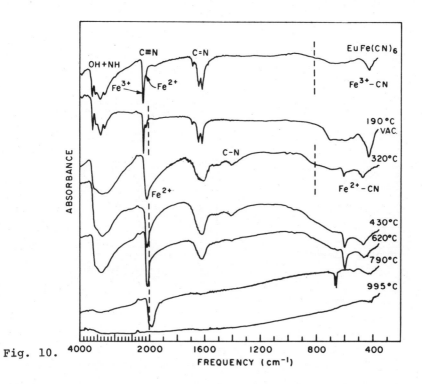

Fig. 10.

Fig.8 Evolved gas analysis of $EuFe(CN)_6 \cdot 5H_2O$
    H_2O ——·· $(CN)_2$ - - - -
    HCN ———   CO and N_2 —·—··—

Fig.9 Evolved gas analysis of $NH_4EuFe(CN)_6 \cdot 4H_2O$
    H_2O ——·· NH_3 - - - -
    HCN ———   CO and N_2 —·—··—

Fig.10 Infrared spectra of $EuFe(CN)_6 \cdot 5H_2O$ heated in vacuum.

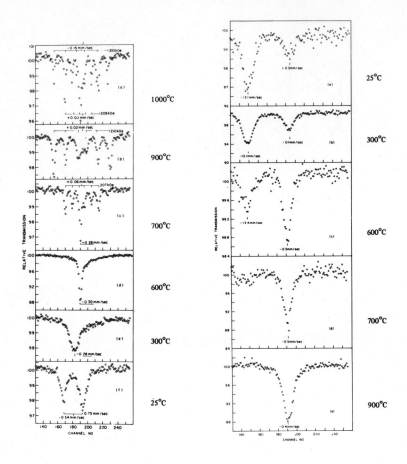

Fig.11(left) ^{57}Fe Mössbauer spectra of EuFe(CN)$_6$·5H$_2$O heated in vacuum.

Fig.12(right) ^{151}Eu Mössbauer spectra of NH$_4$EuFe(CN)$_6$·4H$_2$O heated in vacuum.

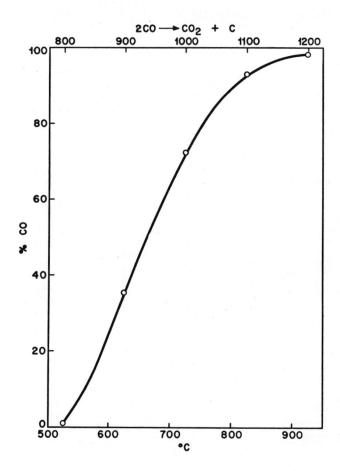

Fig.13 The extent of the disproportionation of CO.

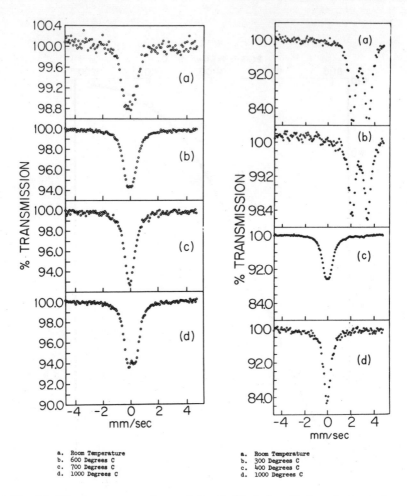

a. Room Temperature
b. 600 Degrees C
c. 700 Degrees C
d. 1000 Degrees C

a. Room Temperature
b. 300 Degrees C
c. 400 Degrees C
d. 1000 Degrees C

Fig.14 Mössbauer spectra of $Ba_2Sn(C_2O_4)_4 \cdot 6H_2O$ calcined in air and measured at room temperature.

Fig.15(right) Mössbauer spectra of $BaSn(C_2O_4) \cdot 1/2\ H_2O$ calcined in air and measured at room temperature.

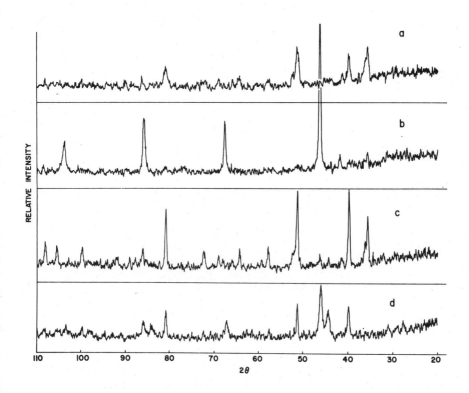

Fig.16 X-ray diffraction patterns using CrK$_\alpha$ radiation.
     a. BaSn(C$_2$O$_4$)$_2$, 30 min. in air at 600° C
     b. BaSn(C$_2$O$_4$)$_2$, 30 min. in air at 700° C
     c. BaCO$_3$·SnO$_2$, 30 min. in air at .900° C
     d. BaCO$_3$·SnO$_2$, 30 min. in air at 1000° C

THERMAL ANALYSIS OF PRECIPITATION PROCESSES IN ALUMINUM ALLOYS

Ken-ichi Hirano

Department of Materials Science
Faculty of Engineering, Tohoku University
Aoba, Sendai 980, Japan

ABSTRACT

Specific heat versus temperature curves have been obtained
by the Nagasaki-Takagi method on polycrystalline specimens of
Al-rich Al-Zn-Mg alloys, during reheating after various age-
hardening treatments. The results have been analysed in comparison with those from hardness tests, X-ray structural and
electrical resistivity measurements, and electron microscopic
observations. The thermal analysis is useful in providing further
information on the low-temperature ageing or the first stage of
ageing, its relation to the high-temperature ageing, formation of
the G.P.zones and metastable phases. The reversion phenomena in
the alloys aged at low temperatures are attributed to the dissolution of the G.P.zones. The effects of the ageing temperature and
ageing time upon both the reversion temperature and the heat of
reversion have been examined from the heat absorption in the
specific heat versus temperature curve, and it has been confirmed
that the reversion temperature rises with ageing temperature.

INTRODUCTION

Precipitation of particles of a second phase from a supersaturated solid solution is responsible for age-hardening in an

alloy system in which the solid solubility decreases with decreasing temperature.

After heating the alloy at a temperature above the solvus for sufficient length of time to dissolve the precipitate, the alloy is quenched into low temperature (usually in water of room temperature) to retain supersaturated state, then reheated to an intermediate temperature at which the age-hardening is possible to occur.
The age-hardening is the most effective method of strengthening aluminum alloys such as commercial Al-Cu-Mg, Al-Zn-Mg and Al-Mg-Si alloys.
It has been shown that the precipitated phase can reach some metastable states prior to achieving equilibrium and these non-equilibrium precipitates are the most effective hardening agents.

By X-ray diffraction measurements, neither new lines for the precipitates nor any change of lattice parameter due to decomposition of the matrix solid solution has been observed during the low temperature ageing (the LT-ageing), whilst these are both clearly observed during the high temperature ageing (the HT-ageing). In 1938, Guinier(1,2) and Preston(3,4) found independently that streaks developed in Laue photograms during the early stage of the LT-ageing in an Al-Cu single crystal, and concluded that within the matrix solid solution there existed Cu-rich clusters of very small size, named the Guinier-Preston Zones(G.P.zones), whose stability and size being dependent upon temperature and time of the LT-ageing. Similar diffraction effects have been observed in other aluminum alloys, The G.P.zones have been also detected by X-ray small angle diffraction measurement and electron microscopy (5,6).

Precipitation sequence in aluminum alloys is usually of the form:
Supersaturated Solid Solution → G.P.Zones (coherent with matrix) → Metastable Phase (partially coherent with matrix) → Equilibrium Phase (incoherent with matrix).

However, it appears that the question whether or not there is any actual relation between the LT-ageing, that is, the creation of the G.P.zones, and the HT-ageing, that is, the precipitation of the metastable and the equilibrium phase, has remained open.

If the alloy is reheated at higher temperature, after an ageing heat treatment sufficiently long for the formation of the G.P.zones, the recovery of hardness is first observed. This recovery is attributed to dispersion of the G.P.zones and called "reversion". But thereafter, hardness-increasing follows as a result of the precipitation of the metastable or equilibrium phase at that temperature.

The formation of the G.P.zones is accompanied with heat evolution, therfore the reversion is to be measured as the absorption of heat in specific heat versus temperature curve obtained during heating the alloy. The calorimetric investigation appears to be more suitable than X-ray measurement and electron microscopy for study especially on thermal stability of the G.P.zones.

The author and his collaborators have investigated the precipitation processes in various aluminum alloys such as Al-Ag, Al-Cu, Al-Zn, Al-Mg-Si and Al-Zn-Mg alloys analysing the specific heat versus temperature curves (7 - 13). In this paper, as representative results, those on an Al-Zn-Mg alloy will be reported and discussed.

EXPERIMENTAL METHODS

The alloy was prepared from 99.99% Al, 99.9% Mg and 99.99% Zn by melting in an alumina crucible under a NaCl-KCl flux and casting into an iron mould. The cast ingots were hot forged and annealed at 470°C for 9 days in order to disperse the casting structure. The grain size of the alloy after these treatments was between 2 mm and 4 mm in diameter.

44

Specimens for the specific heat measurements are closed hollow cylinders, 20 mm in diameter and 30 mm in length as shown in Fig.1. Specimens for the hardness measurements are 1 mm thick sheets prepared by cold rolling; during the cold rolling intermediate annealing was done two times at 500°C for 2 hours. The specimens were homogenized at a temperature above the solvus temperature for 2 hours in an electric furnace and quenched into ice water to obtain the supersaturated state.
The quenched alloys were aged at various temperatures. Ageing below 130°C was carried out in an oil bath regulated within ± 0.5°C by a thermister temperature controller and above 160°C in an electric furnace in air, the temperature being controlled within ± 3°C by a thermocouple controller.

Specific heat versus temperature curves of the aged alloys were obtained using Nagasaki-Takagi type adiabatic calorimeter (14) with an automatic controller and recorders, the details of it being illustrated in Figs.2, 3 and 4.
The curves were obtained during re-heating the specimens from room temperature at a rate of about 1°C per min. All the measurements were carried out in pure argon gas atomosphere (200 mm Hg).

Changes of the hardness of the alloy during ageing and reversion were measured by the Vickers micro-hardness tester at room temperature with 200 gram loading. All the readings are averages of seven impressions.

Heat treatments for the reversion were carried out in a molten nitrate salt bath which was vigorously stirred to homogenize temperature.

PRECIPITATION PROCESSES IN Al-Zn-Mg ALLOY

The Al-Zn-Mg alloy is known as one of the strongest age-hardenable aluminum alloys. Complicated ageing behaviours in this alloy are reported by many workers (6). The precipitation process in a series of Al-Zn-Mg alloys, deduced from the X-ray and electron microscopic investigations, is summarized as follows:

Supersaturated Solid Solution $\rightarrow$ Spherical G.P.Zones $\rightarrow$
Internally Ordered G.P.Zones $\rightarrow$ Metastable M' Phase (or η') $\rightarrow$
Metastable M Phase (or η), $MgZn_2$ $\rightarrow$
Stable (equilibrium) T Phase, $Al_2Mg_3Zn_3$.

The full sequence of the precipitation in the alloy depends upon both alloy composition and ageing temperature.

Age-hardening characteristics in Al-Zn-Mg alloys of various Mg:Zn concentration ratio were studied by Polmear et al. (15), and Morinaga et al. (16). They showed that the G.P.zones were responsible for the initial rise in hardness. The former authors correlated the hardening behaviour with the stability of the G.P. zones. They showed that the stability of the G.P.zones increased with increasing Mg content at a fixed concentration of Zn and vice versa.

Hirano and Takagi (8) discussed the stability of the G.P.zones in an Al-8 wt.%Zn-4 wt.%Mg alloy based on the calorimetric measurements and showed that the stability of the zones increased in proportion to the ageing temperature.

In the present investigation, an attempt has been made to clarify the causes of the complicated precipitation process in the ordinary Al-Zn-Mg alloys. An Al-$MgZn_2$ pseudo-binary alloy system was selected in order to compare the ageing sequence in this alloy with that in Al-$Al_2Mg_3Zn_3$ quasi-binary alloy investigated by Hirano and Takagi (8).

In Fig.5, hardness-ageing time curves of the alloy after quenching from 500°C are shown. When the alloy was aged at room temperature and 40°C, initially the hardness increased almost linearly with the logarithm of ageing time, and subsequently a hardness plateau analogous to that in Al-Cu alloy is observed. On ageing at 70°C, no hardness plateau appeared, but an abrupt change in hardness occurred after 3 hours. On ageing at 160°C, a slight age-hardening was observed. General features of the age-hardeing curves are similar to those of Al-Cu alloy, suggesting tHe precipitation process in this alloy to be similar to that in Al-Cu alloy.

Fig.6 shows the specific heat versus temperature curves (S-T curves) of Al-6 wt.% $MgZn_2$ alloy obtained in the course of re-heating immediately after quenching and slow cooling from 500°C, respectively. The base line, shown by the chain line in the figure, was obtained by the Kopp-Neumann rule from the S-T curves of pure Al, Zn and Mg. The curves in Fig.6 are similar to those of Al-12 wt.% $Al_2Mg_3Zn_3$ alloy obtained by Hirano and Takagi (8). The heat evolution Q between 160° and 230°, of about 45 cal per mole of the disordered homogeneous solid solution, is due to the precipitation of M' and or M phase from the supersaturated matrix during re-heating. The heat absorption R between 230° and 340° is due to re-dissolution of the precipitates.
The amount of R is equal to the heat evolved at Q. The dip in the heat absorption R near 230°C is probable. due to the heat evolution associated with the transformation of the intermediate M' phase into the stable M phase during heating. Similar heat effect was also observed in the transformation of the intermediate θ' phase into the stable θ phase in Al-Cu alloy.

The heat absorption S in the S-T curve of the slowly cooled alloy is about 137 cal per mole. This is required to dissolve the equilibrium M phase into the matrix. The temperature where the heat absorption S ends (340°C) can be regarded as the solvus temperature for this alloy.

The $S-T$ curves of the alloys which were quenched from 500°C
into ice water and aged at 40°, 70°, 100° and 160°C for various
ageing period are shown in Figs. 7, 8, 9 and 10, respectively.
Dashed curve in these figures represents the $S-T$ curve of the
as-quenched specimen.

Similar experiments were carried out on the alloys aged at
room temperature, 130°, 190°, 220° and 250°, respectively.

Ageing at 40°C

As shown in Fig.7, the $S-T$ curves of the alloy aged at 40°C
show the heat absorption P prior to the heat evolution Q.
P starts at a temperature denoted by T_S, passes a maximum at T_M
and finishes off at T_F. Comparing these $S-T$ curves with the
quenched in Fig.6, it is justified that these heat effects represent
the reversion energy, i.e. the energy required for re-dissolution
of the G.P.zones formed during the low temperature ageing.
The heat absorption P increases gradually with ageing time up to
about 3000 hours. On the other hand, the apparent decrease in the
heat evolution, Q, represents the progress in precipitation of the
intermediate M' phase by the ageing. Above about 180°C, the $S-T$
curves of the aged alloy coincide woth the curve (a) in Fig.6 of
the quenched alloy.

Ageing at 70°C

As shown in Fig.8, characteristics of the $S-T$ curves of the
alloy aged at 70°C are essentially similar to those for 40°C-ageing.
The rate of the increase in the heat absorption P and decrease in
the heat evolution Q are rather rapid. After 6 hours of ageing,
the heat absorption P is already observable. The amount of P
increases rapidly until 124 hours and then shows a tendency to
saturate. It should be noted that the shape of the heat absorption curve above 250°C is not different from that of the alloy as
quenched or aged at 40°C.

Ageing at 100°C

The $S-T$ curves for the 100°c-ageing shown in Fig.9 are somewhat different from those for the lower temperature ageing.
As shown in the curve (a) in Fig.9, after ageing for 6 hours, only
a single heat absorption is observed over the lower temperature
range. This heat absorption corresponds to the reversion of the
G.P.zones as in the case of 40°C and 70°C-ageing; however, after
ageing for 63 hours and 112 hours, there appear two heat absorption
peaks as denoted by A and B in the curves (b) and (c).
The heat absorption curve again becomes a single peak after ageing
for 474 hours as shown in the curve (d). These facts indicate
that the second heat absorption B is merely apparent, and both A
and B correspond to each part of a continuous single heat absorption P, because P and Q overlap each other in the common range of
temperature.

The curves (a), (b), (c) and (d) in Fig.9 show that Q decreases
rapidly with the ageing, while R remain nearly constant over a long
ageing interval. It should be noted that the heat absorption P is
considerably large in the curves (d) and (e) in Gig.9.

Ageing at 160°C

Fig.10 shows the $S-T$ curves for 160°C-ageing. At the early
stage of the ageing, it is difficult to separate the heat effect
due to the reversion of the G.P.zones from the whole heat change.
But the curve (d) in Fig.10 clearly shows that the G.P.zones were
formed by ageing at 160°C for 634 hours. In this case the heat
absorptions P and R overlap each other. This feature is also
observed after ageing at 130°, 190°, 220° and 150°C, and is universal when the G.P.zones and an appreciable amount of precipitates
are in co-existence.

Reversion of G.P.zones

In order to verify whether the heat absorption P in Figs.7, 8

9 and 10 was truly responsible for the reversion of the G.P.zones or not, another experiment was carried out. For this purpose, 100°C-ageing was selected because it was found that P is remarkable in this case. Fig.11(a) shows the $S-T$ curve of the alloy aged at 100°C for 783 hours. Dashed curve above T_F (234°C) represents the result obtained from the other specimen which was subjected the same heat treatments. After being heated up to T_F, the specimen was quickly taken out from the calorimeter, quenched into water at room temperature and then the specific heat was measured again. The result is shown in Fig.11(b) where the heat absorption P disappeared completely and in place of P, a heat evolution Q was observed at about 200°C. This heat evolution may correspond to that obseved on the specimen immediately after quenching, Fig.6(a). This means that at the heat absorption P in Fig.11(a), the solute atoms return from the G.P.zones to the matrix and the partly super-saturated state is reproduced in the matrix at the end of the heat absorption P. The results in Fig.11 support the interpretation that P represents the re-dissolution of the G.P.zones.

In Figs.7 - 10, the reversion of the G.P.zones starts at T_S, becomes most considerable at T_M and finishes off at T_F. These temperatures can be taken as measures of the thermal stability of the G.P.zones.

Fig.12 shows the ageing time dependence of T_M at 40°, 70° and 100°C. In the alloy aged at 100°C, T_M increases initially with ageing time and then shows a tendency of saturation. In the alloys aged at 40° and 70°C, T_M increases gradually and almost linearly with the logarithm of ageing time, and does not saturate even after 3000 and 2000 hours, respectively. T_M may saturate after a pro-longed ageing at these temperatures. An essentially similar behav-iour of T_M has been observed in Al-Cu, Al-Ag, Al-Zn and Al-Al$_2$Mg$_3$Zn$_3$ alloys.

Fig.13 shows the relationship between T_S, T_M, and T_F and the ageing temperature, T_A. These reversion temperatures were determined for the alloy aged sufficiently at each temperature after quenching from 500°C. T_S increases almost linearly with ageing temperature. The line for T_S is parallel to the line corresponding to $T_S = T_A$. T_S is higher by about 35° than T_A, showing that the reversion of the G.P.zones begins at a temperature higher than the ageing temperature by 35°C.

T_M and T_F increase linearly with T_A between room temperature and 160°C, and above 190°C they deviate from the linear relationship. T_S, T_M and T_F seem to coincide at the solvus temperature, T_P, when they are extrapolated as shown in Fig.13. It should be noted that T_P also coincides with the solvus temperature of the equilibrium $MgZn_2$ phase in this alloy.

As shown in Figs.12 and 13, the reversion temperatures depend upon both the ageing time and temperature. The variation in T_M with ageing time at a constant ageing temperature may be correlated to the change in the stability of the G.P.zones. Comparing Fig.9 with Fig.12, it is found that the reversion of the G.P.zones occurs in the same temperature range regardless of the solute concentration retained in the matrix. This fact shows that the stability of the G.P.zones is not so sensitive to the matrix concentration during the ageing. In Al-Cu alloys, it has been pointed out that the zone size plays an important role in the stability (10). Increasing of the stability of the zones with size may be related to the Thomson-Freundlich formula for the variation of solubility with particle size. When the particle is small, the chemical potential of the solute atom in the particle is high because of its surface tension and the solute has a tendency to leave from the particle. As the particle becomes larger, the effect of the interfacial energy on the solubility becomes negligible and the particle is stabilized. By some calculations, it can be found that the effect of the particle radius on the solubility is remarkable for the particle smaller than 50 A.

Kinetic analysis of S-T curve to
determine activation energy for reversion

According to Nagasaki et al. (17), it is possible to analyse kinetically the $S-T$ curve under the condition of a constant heating rate. Applying their method to the reversion peak P and assuming the first order reaction for the dissolution of the G.P.zones, the following equation is obtained:

$$\ln \{(C_m - C_b)/C_b\} - \ln (P_O - P_T) = \ln A - E/RT$$

where C_m, C_b, P_O and P_T are defined as shown in Fig.14, A is a constant, E the activation energy for the reaction, T the absolute temperature and R the gas constant.

In Fig.15, $\ln[\{(C_m - C_b)/C_b\}/(P_O - P_T)]$ is plotted against the reciprocal of the absolute temperature for the $S-T$ curves of the alloys aged at room temperature, 40°, 70° and 100°C, respectively. Straight lines, ab , cd , ef and gh , correspond to the reversion of the zones and three straight lines, ij , kl and mn , correspond to the precipitation of the M' or M phase.

It can be seen that for the alloy aged at 100°C, only single straight line is obtained, suggesting that the heat absorption P in the alloy aged at 100°C for a long time consists of only one reaction of the first order, i.e. purely the reversion.

From the slope of the line, the activation energy for the reversion was obtained as indicated in Fig.15. It should be noted that the activation energy is not constant, but varies with ageing temperature, from 15 kcal/mol for the alloy aged at room temperature to 25 kcal/mol for the alloy aged at 100°C. The lower activation energy for the reversion than that for the solute diffusion of Zn or Mg in Al or the interdiffusion in Al-Mg and Al-Zn couples may be correlated to the behaviour of the quenched-in vacancies retained in the matrix and the G.P.zones.

Ageing characteristics of Al-6 wt.% MgZn$_2$ alloy

The energy difference between aged and quenched alloys, which is given by the hatched area in Figs.7 - 10, corresponds to the heat evolved during the ageing and is a measure of the progress in the precipitation. This energy difference is plotted against the ageing temperature in Fig.16 by the closed circles. On the other hand, the heat absorption P is correlated to the amount of the G.P. zones. This energy change is shown by the open circles in Fig.16 (curve I). The difference between the total energy change and the heat absorption P at each ageing temperature may be attributed to the precipitation of the M' or M phase. This is shown by the curve II, which appears to consists of two parts, A and B, as shown in Fig.16.

As shown previously, the different shapes of the heat absorption curve, R, in the S-T curves were observed between the alloys aged below 70° and aged above 100°C. Since the dip in R in the alloys aged at 40° and 70°C has been interpreted to be due to the transformation of the M' into the M phase, it is likely that the M' and M phases are the main precipitates when the alloy is aged below 70° and above 100°C, respectively.
The curves A and B in Fig.16 can be attributed to the heat effect due to the M' and M phase, respectively.

The total energy change in Fig.16 increases first with ageing temperature, passes a peak at about 130°C, then decreases nearly linearly and diminishes at about 340°C which corresponds to the solvus temperature, T_p. It is clear that in the lower temperature range, the formation of the G.P.zones causes a rapid energy change. The peak at 130°C can be attributed to the co-existence of the G.P. zones, the M' and M phases.

Although the distinction between the LT-ageing and HT-ageing is clearly shown in Fig.16, the G.P.zones are formed even in the

HT-ageing range. However, the amount of the zones is small in
the HT-ageing range and the contribution of them to age-hardening is negligible. It is likely that the zones could not be
detected by electron microscopy because of the mutual interference
of a large amount of the intermediate and equilibrium precipitates
in co-existence.

The fact that no heat evolution was observed in Fig.9(d) and
(e), indicates that all the supersaturated solute atoms had been
already condensed into the G.P.zones and the precipitates before
474 hours at 100°C. The presence of the heat absorption P under
such condition implied that the G.P.zones can coexist with the
equilibrium precipitates even after a prolonged ageing.
This suggests that there is a certain metastable equilibrium with
respect to the G.P.zones, and a long ageing is required to obtain
the final equilibrium state.

CONCLUSIONS

Ageing characteristics in a supersaturated Al-6 wt.% $MgZn_2$
quasi-binary alloy have been investigated principally by a calorimetric method with particular emphasis on the thermal stability of
the Guinier-Preston zones. Ageing sequence in the alloy is as
follows; Supersaturated Solid Solution → G.P.zones → Intermediate
M' phase → Satble M phase ($MgZn_2$). It has been found that the
G.P.zones can be formed on ageing below 340°C which corresponds to
the solvus temperature of the M phase. The intermediate phase M'
is formed preferentially on ageing below 70°C and the equilibrium
M phase precipitates predominantly above 100°C.

Reversion temperature of the G.P.zones increases with ageing
time for a certain period, after which it has a tendency to saturate
to a definite temperature depending on the ageing temperature.
The reversion temperature varies from 170°C for the alloy aged at
room temperature to 300°C for the alloy aged at 190°C.

It has been found that the activation energy for the reversion of the G.P.zones determined from the specific heat versus
temperature curve depends upon the ageing temperature.

In general, it hàs been shown that the thermal analysis
is very useful to investigate the precipitation reaction in super-
saturaṭed solid solution.

REFERENCES

1. A.Guinier, Compt.rend. 206, 1641 (1938)
2. A.Guinier, Nature 142, 569 (1938)
3. G.D.Preston, Proc.Roy.Soc. A167, 536 (1938)
4. G.D.Preston, Phil.Mag. 26, 855 (1938)
5. H.K.Hardy and T.J.Heal, Progress in Metal Physics, Ed. by
   B.Chalmers and R.King, Vol.5, 143 (1954)
6. A.Kelly and R.B.Nicholson, Progress in Materials Science, Ed.
   by B.Chalmers, Vol.10, No.3 (1963)
7. K.Hirano, J.Phys.Soc.Japan 8, 603 (1953)
8. K.Hirano and Y.Takagi, J.Phys.Soc.Japan 10, 187 (1955)
9. K.Hirano, J.Phys.Soc.Japan 10, 995 (1955)
10. K.Hirano and J.Iwasaki, Trans.Japan Inst.Metals 5, 162 (1964)
11. K.Hirano, Trans.Japan Inst.Metals 10, 132 (1969)
12. K.Asano and K.Hirano, Trans.Japan Inst.Metals 9, 24 (1968)
13. T.Miyauchi,S.Fujikawa and K.Hirano, Keikinzoku (Light Metals)
    21, 565 (1971)
14. S.Nagasaki and Y.Takagi, J.Appl.Phys.Japan 17, 104 (1948)
15. I.J.Polmear, J.Inst.Metals 89, 51 (1960)
16. T.Morinaga,T.Takahashi,J.Yamashita and G.Yokota, Keikinzoku
    (Light Metals) 15, 275 (1965)
17. S.Nagasaki and A.Maesono, Kinzoku Butsuri (Metal Physics) 11,
    182 (1965)

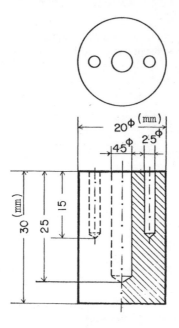

Fig.1   Sample for Nagasaki-Takagi type
         adiabatic calorimeter

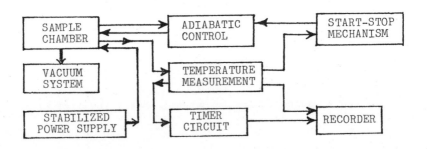

Fig.2   Schematic diagram of Nagasaki-
         Takagi type adiabatic calorimeter

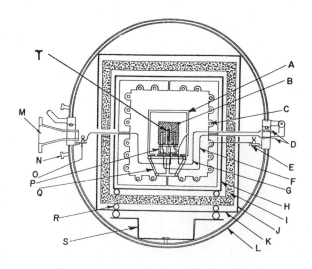

Geometrical Arrangement illustrating Nagasaki-Takagi Type Adiabatic Calorimeter.

A: Adiabatic container (pure Ni)
B: Sample
C: Electric heater (outside heater)
F: Thermocouple for temperature measurement
G: Thermocouple for adiabatic control
H, J and K: Shielding plate (stainless steel)
I: Shielding plate (ceramics)
L: Container
P, Q and R: Quartz supports
S: Support (stainless steel)
T: Electric heater (internal heater)

Fig.3 Nagasaki-Takagi type adiabatic calorimeter

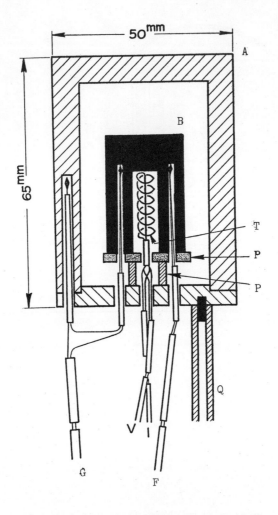

Fig.4 Adiabatic specimen container of Nagasaki-
Takagi type calorimeter

58

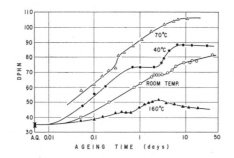

Fig.5   Hardness-time curves of Al-6 wt.% MgZn$_2$
         alloy aged at room temperature, 40°,
         70° and 160°C after quenching from 500°C

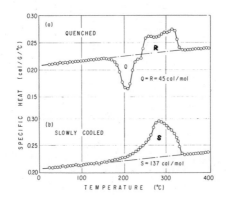

Fig.6   Specific heat versus temperature curves of
         Al-6 wt.% MgZn$_2$ alloy
         (a) quenched   (b) slowly cooled

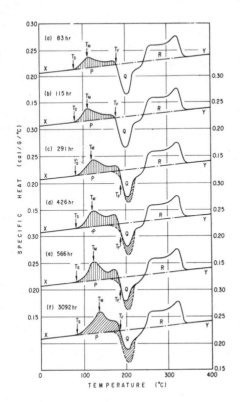

Fig.7   Specific heat versus
         temperature curves of
         Al-6 wt.% MgZn$_2$ alloy
         aged at 40°C

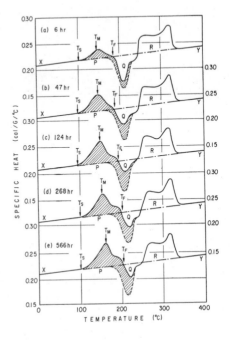

Fig.8   Specific heat versus
         temperature curves of
         Al-6 wt.% MgZn$_2$ alloy
         aged at 70°C

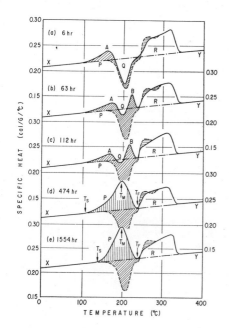

Fig.9   Specific heat versus
         temperature curves of
         Al-6 wt.% MgZn$_2$ alloy
         aged at 100°C

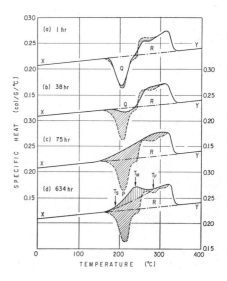

Fig.10   Specific heat versus
         temperature curves of
         Al-6 wt.% MgZn$_2$ alloy
         aged at 160°C

61

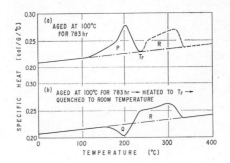

Fig.11   Specific heat versus temperature curves of
          Al-6 wt.% MgZn$_2$ alloy
          (a) aged at 100°C for 783 hours
          (b) aged at 100°C for 783 hours, then heated
              to T$_F$ (234°C) and quenched into water
              at room temperature

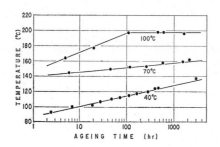

Fig.12   Change of T$_M$ of Al-6 wt.% MgZn$_2$ alloy
          during ageing at 40°,70° and 100°C.

62

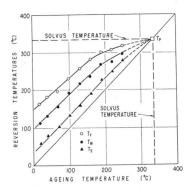

Fig.13   Ageing temperature dependence of the
          reversion temperatures of Al-6 wt.%
          MgZn$_2$ alloy

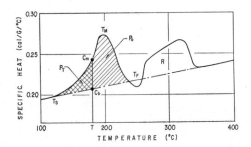

Fig.14   Illustration of the kinetic analysis of
          the specific heat versus temperature
          curve

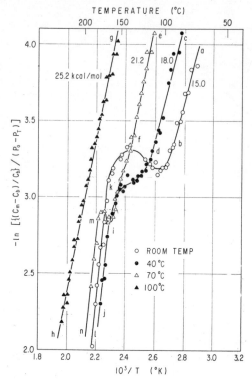

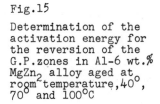

Fig.15

Determination of the activation energy for the reversion of the G.P.zones in Al-6 wt.% MgZn$_2$ alloy aged at room temperature, 40°, 70° and 100°C

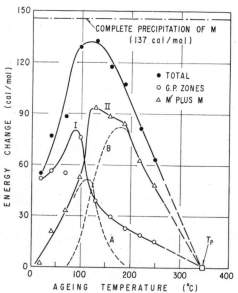

Fig.16

Energy difference between the aged and as-quenched alloys as a function of the ageing temperature

THERMAL ANALYSIS IN SEALED TUBES:
INORGANIC, ORGANIC AND BIOCHEMICAL REACTIONS

Horst W. Hoyer

Hunter College of the City University of New York

ABSTRACT

The use of sealed glass vials in differential thermal techniques permits study of chemical systems at temperatures of up to 550°C and at pressures of up to 50 atmospheres. Among the reactions which have been
studied are cis-trans isomerizations, Diels-Alder reactions, polymerizations, critical temperature determinations, high energy decompositions and
the helix-random coil transition of deoxyribonucleic acid. The technique
permits rapid evaluation of the effectiveness of catalysts for chemical reactions.

INTRODUCTION

During the greater part of its history differential thermal analysis
has been used extensively for the characterization of inorganic and organic
substances. In recent years it has also developed into an important technique for studying chemical reactions. Conventional DTA techniques are
difficult to apply to such systems for a number of reasons. Vaporization
and sublimation of the material or evaporation of the solvent may take
place before the reaction gets under way or the reaction may occur above
the boiling point of the sample or solvent. Such processes produce extra-
neous thermal effects which distort that produced by the reaction under investigation.

An obvious solution is to enclose the system in a sealed container
which will minimize evaporation of solvent and/or sample. However, the conventional crimping technique by which samples are sealed in aluminum vials
does not permit operation much above the boiling point of the solvent or
sample. We have found that small glass vials and ordinary melting point
capillaries are suitable containers for these reactions. When properly
sealed they can withstand pressures to 55 atmospheres. However we normally
try to limit our sample size so that the internal pressure does not exceed
45 atmospheres during the experiment because, while explosions of the vials
and capillaries in the thermal analysis apparatus are not serious, they do
eventually damage the thermocouples.

The use of this sealed cell technique has made it possible to study a
wide variety of chemical and physical reactions occurring at temperatures of
up to about 550°C and to pressures of about 50 atmospheres. Those studied
in our laboratory have included cis-trans isomerizations, polymerization reactions, Diels-Alder reactions, decomposition of high energy compounds,
evaluation of the effectiveness of catalysts, estimation of critical temperatures and the helix-coil transition of deoxyribonuclic acid (DNA).

65

DESCRIPTION OF TECHNIQUE

Our initial work with the sealed cell technique was carried out with the calorimeter cell of the duPont differential thermal analyzer shown in Figure 1. The glass reaction vials, available from the duPont Company, are 4 millimeters in diameter and 28 millimeters in length. Small known quantities of material, from 50 to 100 milligrams, are sealed in the vials with the aid of a special low temperature jig shown in figure 2. Later, when the differential scanning calorimeter cell became available for the duPont analyzer we used ordinary melting point capillaries in which were sealed 2 to 4 mg. of sample.

The jig, a rectangular block of aluminum approximately 10 cm x 2 cm x 6 cm, has several holes drilled into the top to accommodate vials and capillaries. The holes for the vials are approximately 1 cm. deep, those for the capillaries are 0.4 cm. deep. These depths are dictated by the dimensions of the apparatus in which the thermal analysis is run and may differ from one type of apparatus to another.

In order to secure good thermal contact between the vials and the walls of the jig, a drop of water is placed into the well before the vial or capillary is inserted into the opening. The temperature of the jig and sample is then reduced by immersing the lower part of the jig in a dry ice-acetone bath. After several minutes the jig is removed from the bath and the vial or capillary immediately sealed with a micro torch. Care should be taken to avoid the formation of microscopic pinholes during the sealing process since escape of sample or solvent during the experiment will mitigate the results. When melting point capillaries are used, care must be taken to avoid contamination of the sides of the tube since this leads to difficulty in sealing. With the vials the greatest source of difficulty in securing a tight seal seems to be slowness in sealing. With practice and speed, one can consistently seal the vials and capillaries to withstand pressures of up to 55 atmospheres.

Often it is desirable to exclude oxygen from the reaction vial or to carry out the reaction in a special inert atmosphere. At other times it is convenient to distill the sample directly into the reaction vial. For such cases we have designed and used an all-glass distillation tree which was constructed by the local glassblower and is shown in Figure 3. The samples are distilled directly into the vials, either under vacuum or in an inert atmosphere, and the vials then sealed and removed from the tree. During the sealing process, it is essential that the sample and vial be kept cold with a small amount of dry ice. Otherwise the high vapor pressure generated by the heat prevents adequate sealing.

Much of our initial work involved the 4 mm vials. Now, however, we prefer the capillary tube method because the tubes require less sample and are cheaper and more readily available than the vials. Thermograms by either method appear similar in every respect. The vials do have one important advantage over the capillary tubes in that the larger sample size, from 50 to 100 mg, is more than adequate for sample analysis by modern techniques such as gas chromatography or infrared spectroscopy and makes it possible to follow the sequence of chemical reactions occurring during a thermogram. However, should the internal pressures exceed about 55 atmospheres, the explosions of the vials may damage the calorimeter cups of the DTA apparatus. Explosions with the capillary tubes in the DSC apparatus are not as damaging. Normally we try to prevent explosions by limiting the sample size to that which could generate a pressure of no more than 45 atmospheres. It should be pointed out that after an experiment the capillaries and vials may be under considerable internal pressure. Goggles or protective glasses should be worn when removing the vials or capillaries from the apparatus and they should be crushed behind a shield or under an adequate volume of water.

Two of the cis-trans isomerization reactions which we have studied by DTA are the cis-trans rearrangement of stilbene[1] and that of oleic acid to its trans isomer, elaidic acid.[2] Figure 4 shows the thermograms obtained on heating cis-stilbene in a sealed vial. At approximately 350°C, with a vapor pressure of about 40 atmospheres, the cis form spontaneously rearranges to the trans form. Figure 5 shows a similar type of rearrangement for cis-oleic acid to its trans isomer, elaidic acid. In this case the reaction occurs at a much higher temperature and is accompanied by decomposition of the sample. Both thermograms show the marked effect of catalysts on the reactions. Indeed one obvious advantage of DTA to the study of chemical reactions is the ease with which it permits evaluation of the effectiveness of various catalysts.

Figure 6 shows results we obtained during a thermal study of the polymerization of styrene.[3] There is little difference between the bulk polymerization and polymerization in toluene solution; polymerization being initiated at about 200°C. However, the effect of the benzoyl peroxide catalyst is quite pronounced.

One of the interesting reactions which can be unraveled by this technique is the Diels-Alder dimerization of cyclopentadiene.[4] Pure cyclopentadiene shows three distinct exothermic peaks during the DTA experiment, Figure 7(A). The thermogram for pure dicyclopentadiene, 7(B), shows only two exothermic peaks but in addition has an endotherm preceding the two peaks. Samples of the cyclopentadiene and dicyclopentadiene were then run and quickly removed from the DTA apparatus at temperatures corresponding to the various exothermic and endothermic peaks. These samples were analyzed by infrared spectroscopy and gas chromatography. Peak (I), as suggested by comparison of thermograms A and B, corresponds to the formation of the endodicyclopentadiene. The endotherm located between (I) and (II) is accompanied by the reappearance of the cyclopentadiene due to the depolymerization of the endo-dimer. This is followed in region (II) by the appearance of the exo-dicyclopentadiene. Apparently the rearrangement of dicyclopentadiene from the endo to the exo form involves the initial decomposition of the endo form. In Region III we observed the appearance of a black tarry mass indicating polymerization accompanied by decomposition.

Even high energy compounds which are normally difficult to study and dangerous to handle can be investigated by this technique.[5] Figure 8 shows the rapid decomposition of dimethyl hydrazine at about 400°C during a DTA experiment.

During the course of our investigations with this sealed cell technique, we discovered that it could be used to determine the critical temperature of liquids[6] provided that their critical pressures were less than the bursting pressure of our vials, approximately 55 atmosphere. The method has permitted determination of the critical temperatures of the seven compounds listed in Table 1 with an average deviation from the literature values of $\pm 0.6^\circ$.

TABLE 1

Comparison of Experimental Critical
Temperatures with Literature Values

Compd.	Exptl[a]	Lit.[b]
Ethyl ether	192.5 + 0.5	193.8
Acetone	234.5 ∓ 0.5	235.0
Cyclohexane	281.5 ∓ 0.5	281.0
n-Hexane	235.6 + 0.5	234.8
n-Heptane	267.0 ∓ 0.5	267.1
Ethyl acetate	248.6 ∓ 0.5	250.1
Benzene	289.4 ∓ 0.5	288.9

a) Corrected for thermal lag using the melting point of p-chlorocinnamic acid as a standard. b) "Handbook of Chemistry and Physics." Chemical Rubber Publishing Co., Cleveland, Ohio, 41st ed. 1959-1960.

Figure 9 shows the thermogram which results when a sample of ethyl ether is cooled from about 240°C in one of the sealed vials. Our explanation of this thermogram involves Figure 10 in which (a) is the density vs. temperature coexistence curve for a system consisting of a pure liquid and its vapor and in which (b) is the thermogram which one might obtain according to the following analysis.

Referring to Figure 10, let us assume that a gas is above its critical temperature and has an average overall density D. Assume further that the density gradient which will develop within the vessel near the critical point is defined by the density limits D_1 and D_2. As long as the range $D_1 - D_2$ includes the critical density, D_c, it is apparent from the figure that cooling the gas must result in some phase separation when the critical temperature is reached. This phase separation is accompanied by an increase in the heat capacity of the system since a two-phase system at constant volume has a larger heat capacity than the gas phase from which it is condensed.

Thus, the corresponding thermogram must show a sharp change at T_c toward a higher heat capacity. This transition continues as the system is cooled through its inhomogeneous region until T_1 is reached. At this temperature the system is once more homogeneous and the thermogram has reached its lower base line. Figure 9, the experimental cooling curve for diethyl ether, is typical of the cooling curves obtained with other samples and illustrates the close agreement between Figure 10 (lower) and an actual DTA cooling curve.

The critical temperatures were measured by obtaining cooling thermograms of the samples in sealed 4-mm glass tubes using a calorimeter cell of a duPont Model 900 differential thermal analyzer. Samples of 20 - 50 μl were added directly into the sample tube and the tube was sealed and placed into the reference reservoir of the calorimeter cell. An open glass tube filled with glass beads and placed in the sample reservoir was employed as a reference. The reversal of sample and reference positions is necessitated by the design of the duPont calorimeter cell which records the reference-compartment temperature on the y axis rather than the sample temperature. The cell was then heated to a temperature above the coexistence curve as evidenced by a rapid change in ΔT $[\Delta T$ (differential temperature) = T(ref) - T(sample)$]$. Heating was then terminated and the cell was allowed to cool while recording ΔT as a function of sample temperature. A sharp break in the curve corresponds to a point on the coexistence curve. The temperature corresponding to the coexistence point was determined as a function of increasing sample volume until a constant temperature (critical temperature) was achieved.

STUDIES ON DNA SOLUTIONS

Although several investigators have reported calorimetric studies on the helix-coil transition of solutions of deoxyribonucleic acid (DNA), the differential thermal studies seem to have been those of Shaio and Sturtivant[7] and of Privalov and coworkers.[8] Our initial attempts to apply the sealed vial technique to this system were troubled by the need to use the extreme sensitivity of our instrument with resulting difficulty in controlling the baseline. Baseline control had been excellent with the other systems which had been investigated and we ascribe the problem to the much lower heat effects accompanying the helix-coil transition, approximately 9 Kcalories per mole base pair whereas for the chemical reactions the values run from 50 to 90 Kcalories per mole.

Sensitivity and instrument response increased considerably when we switched to the DSC apparatus and used the conventional aluminum cups supplied by the manufacturer. These may be crimped shut to withstand an internal pressure of about 2 atmospheres, corresponding to a temperature of about 120°C for aqueous systems.

In the concentrations of about 4 to 6 weight percent of calf-thymus DNA which we were obliged to use, the solutions are viscous gels. The onset of the endothermic transition occurs at approximately 95°C, consistently higher than that reported in more dilute solutions, and is not completed until the temperature reaches about 110°C.[10] A typical endotherm is shown in Figure 11.

Studies were made in solutions of phosphate buffers over the pH range from 5.4 to 8.2. Our results are tabulated in Table 2 and show a maximum in the enthalphy of the helix-coil transition of 9.4 calories per mole base pair at a pH of about 7.0.

TABLE 2

ENTHALPY OF HELIX-COIL TRANSITION IN DNA GELS

pH (Phosphate buffer)	Kcalories per mole base pair (618 g per base pair)
5.4	7.7 $\pm$ 0.3
5.8	7.7 $\mp$ 0.4
6.2	8.5 $\mp$ 0.4
6.6	8.6 $\mp$ 0.2
7.0	9.4 $\mp$ 0.2
7.4	8.7 $\mp$ 0.4
7.8	8.5 $\mp$ 0.5
8.2	7.6 $\pm$ 0.4

Because the differential scanning calorimeter measures the rate of change of enthalphy as a function of time and temperature it permits determination of kinetic as well as thermal data. We have applied the method of Borchardt and Daniels[9] to our results to obtain information concerning the order and activation energy of the helix-coil transition of calf-thymus DNA over the pH range of 5.4 to 8.2.

Figures 12, 13 and 14 show conventional Arrhenius plots at a pH of 7.0 for assumed orders of reaction of zero, first, second and third order. The best straight line, from approximately 10% conversion to 90% conversion is obtained with the second order plot. At high conversions errors in reading the thermograms become significant and deviations from linearity are to be expected. However the initial slope is not as susceptible to such error and is, we believe, significant. It has been suggested that the initial stage of the helix-coil transition of DNA is an initiation process different from the subsequent unwinding and our experiments would seem to support this view. Our values for the activation energies are tabulated in Table 3 and suggest that the activation energy of the unwind-

TABLE 3

ACTIVATION ENERGIES FOR HELIX-COIL TRANSITION IN DNA GELS
(Kcalories per mole kinetic unit)

pH	Second Order E_a Initial 10%	Main Portion	First Order E_a Initial 10%
5.4	177	68	167
5.8	175	68	185
6.2	169	66	167
6.6	190	74	181
7.0	174	65	241
7.4	197	72	206
7.8	192	71	243
8.2	220	71	217
Average	187 $\pm$ 14	69 $\pm$ 3	201 $\pm$ 25

ing process is about 69 Kcalories. The value for the initiation step is more uncertain and would seem to lie between 180 and 200 Kcalories.

REFERENCES

1. A.V. Santoro, E.J. Barrett and H.W. Hoyer, J. Am. Chem. Soc., 89 (1967) 4545
2. E.J. Barrett, H.W. Hoyer and A.V. Santoro, Tetrahedron Letters, 5 (1968) 603
3. H.W. Hoyer, A.V. Santoro and E.J. Barrett, J. Polymer. Sci., Part A-1, 6 (1968) 1033.
4. A.V. Santoro, E.J. Barrett and H.W. Hoyer, Tetrahedron Letters, 19 (1968) 2297.
5. E.J. Barrett, H.W. Hoyer and A.V. Santoro, Anal. Lett., 285 (1968).
6. H.W. Hoyer, A.V. Santoro and E.J. Barrett, J. Phys. Chem., 72 (1968), 4312.
7. D.D.F. Shaio and J.M. Sturtevant, Biopolymers, 12 (1973), 1829.
8. P.L. Privalov, O.B. Ptitsyn, T.N. Birshtein, Biopolymers, 8 (1969), 559.
9. H.J. Borchardt and F. Daniels, J. Am. Chem. Soc., 79 (1957), 41.
10. H.W. Hoyer and S. Nevin, in "Analytical Calorimetry, Volume 3, Plenum Press (1974).

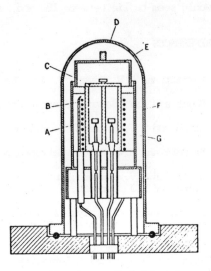

Fig. 1. DuPont calorimeter cell. A: sample holder; B; control thermocouple;
C: heating block cover; D: bell jar; E: insulating lid, F: heater;
G: heating block

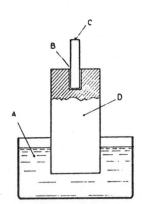

Fig. 2. Jig for sealing glass cells. A: dry ice-acetone bath; B: vial sealed
at that point, C: sample vial; D: aluminum block

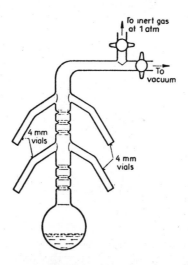

Fig. 3.   Glass distillation apparatus

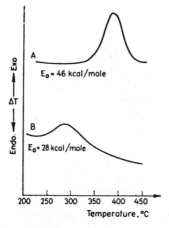

Fig. 4.   Cis to trans isomerization of stilbene.   A: uncatalyzed stilbene; B:
Pd catalyzed stilbene

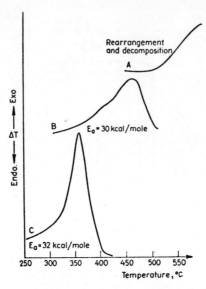

Fig. 5. Cis to trans isomerization of oleic acid to elaidic acid. A: uncatalyzed oleic acid: B: Se catalyzed oleic acid; C: iodine catalyzed
oleic acid

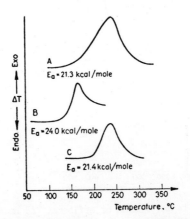

Fig. 6. Polymerization of styrene. A: uncatalyzed styrene; B: benzoyl
peroxide catalyzed styrene; C: styrene in toluene solution

74

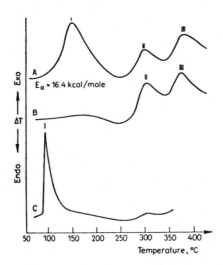

Fig. 7. The Diels-Alder reaction. A: uncatalyzed cyclopentadiene; B: dicyclopentadiene; C: HCl catalyzed cyclopentadiene. Reactions: I: dimerization of cyclopentadiene; II: formation of polymer and conversion of endo-dicyclopentadiene to exo-dicyclopentadiene: III: decomposition and further polymerization

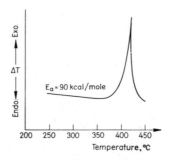

Fig. 8. Decomposition of dimethyl hydrazine

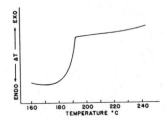

Fig. 9.   Experimental cooling thermogram for ethyl ether

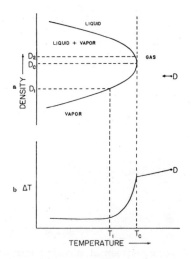

Fig. 10.   (a) Density-coexistence curve; (b) cooling thermogram

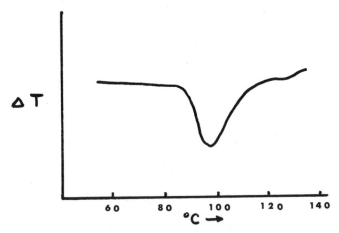

Fig. 11.   Thermogram of DNA in Phosphate buffer.   Buffer concentration is 0.75 moles per liter, pH = 7.0

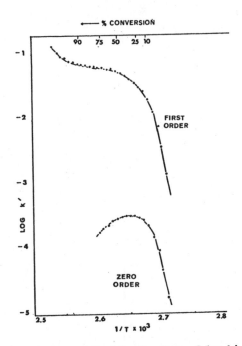

Fig. 12.   Zero and first order kinetic plots of log k' from equation 2 versus reciprocal of absolute temperature for 6.50% DNA in 0.75M phosphate buffer at pH 7.0.

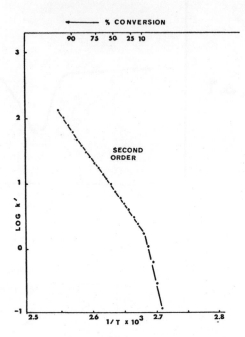

Fig. 13.

Second order kinetic plots of log k'
from equation 2 versus reciprocal of
absolute temperature for 6.50% DNA
in 0.75M phosphate buffer at pH 7.0.
To convert k' to k multiply by 0.12.

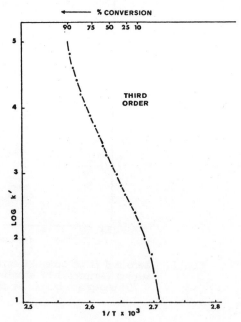

Fig. 14.

Third order kinetic plots of log k'
from equation 2 versus reciprocal
of absolute temperature for 6.50% DNA
in 0.75M phosphate buffer at pH 7.0.

THERMOMECHANICAL INVESTIGATION OF THERMAL SHRINKAGE
OF THE COLD-DRAWN AROMATIC POLYMER FILMS

Hirotaro Kambe, M. Kochi, T. Kato, and M. Murakami

Institute of Space and Aeronautical Science
University of Tokyo
Komaba, Meguro-ku, Tokyo, Japan

ABSTRACT

Thermal shrinkage of the several aromatic polymer films
stretched at the room temperature was analyzed by the dynamic
thermoanalytical method. The decrease of the sample length was
recorded against temperature at various rates of heating, and
the temperatures at the definite contraction levels for various
heating rates were measured. The plots of logarithmic heating,
rate vs. temperature give the activation energy ΔE of the
shrinkage reactions, assuming the first-order reaction.

The method was applied to polypyromellitimide (PI), polytrimellitamideimide (PAI), and poly(Bisphenol-A carbonate) (PC).
From the analysis of thermal shrinkage of stretched PI films,
we observed three processes appeared at 100°C, 250°C, and 350°C,
for which we obtained ΔE's of 10 kcal/mole, 25 kcal/mole, and
35 kcal/mole. From the data obtained for PAI film, we found processes at 80°C, 150°C, and 250°C. The process observed at the
highest temperature corresponds to the glass transition of PAI.
The ΔE's estimated for the lower two processes are 13 kcal/mole
and 17 kcal/mole, respectively. For PC we obtained two processes
at 25°C and 80°C in the glassy state. We also obtained ΔE of
17.4 kcal/mole for the process at higher temperature.

EXPERIMENTAL

Materials   All samples are commercial products. Polypyromellitimide (PI) film is Du Pont Kapton H.

79

Polytrimellitamideimide (PAI) film

and polybisphenol-A carbonate (PC) film

were supplied by Teijin Co., Ltd.

Apparatus and Procedure The specimens were stretched by a tensile machine with a constant speed of stretching at ambient temperatures. The thermomechanical measurements were carried out by the modified Rigaku Denki linear expansion apparatus with the electronic derivative circuit. The sample film was cut to a thin strip of 5 mm in width and 2 to 20 mm in length. The change of length of the specimen was recorded in vacuum against temperature under a uniform heating rate of 10°C/min, under the tensile load of 10 mg. The creep of the film was negligible.

THEORETICAL

The theory of thermoanalytical treatment of contraction has been published in the previous paper(1). A two-state model of the energy states of polymer chain is assumed. In the as-received film, segments of the polymer are assumed to take contracted α or extended β state randomly. At an extended state, the number of segments in β state should be increased. By the increase of temperature, a fraction of segments in β state would move to the α state. We take the number fraction of segments moved from the upper (β) to the lower (α) state at time t as x, and the rate of contraction is assumed to follow the first-order reaction.

Assuming the Arrhenius equation we obtain

$$\frac{dx}{dt} = A \exp\{-\frac{\Delta E}{RT}\}(1 - x) \tag{1}$$

Integrating

$$\int_0^x \frac{dx}{1 - x} = A\int_0^t \exp\{-\frac{\Delta E}{RT}\}dT \tag{2}$$

At a uniform rate of heating, $\frac{dT}{dt} = \phi$

$$-\ln(1 - x) = \frac{A}{\phi}\int_0^T \exp\{-\frac{\Delta E}{RT}\}dT \tag{3}$$

Let $y = \frac{\Delta E}{RT}$ and integrating by parts

$$- \ln(1 - x) = \frac{A\Delta E}{\phi R}\left[\frac{\exp(-y)}{y} + E_i(-y)\right] = \frac{A\Delta E}{\phi R} P(y) \quad (4)$$

where $E_i(-y)$ is the exponential integral

$$E_i(-y) = -\int_y^\infty \frac{\exp(-y)}{y}\, dy \quad (5)$$

The Eq.(4) is very familiar in the kinetic analysis of thermo-gravimetry, and we solve it by Ozawa's method(2). If we can measure the thermal contraction curves at various heating rates ϕ_1, ϕ_2 etc., corresponding temperatures T_1, T_2 etc. are obtained at a certain degree of contraction or x. Then we have

$$\frac{A\Delta E P(\frac{\Delta E}{RT_1})}{\phi_1{}^R} = \frac{A\Delta E P(\frac{\Delta E}{RT_2})}{\phi_2{}^R} = \cdots \quad (6)$$

Assuming Doyle's approximation(3)

$$\log \phi_1 + 0.4567\frac{\Delta E}{RT_1} = \log \phi_2 + 0.4567\frac{\Delta E}{RT_2} = \cdots \quad (7)$$

Then from the linear plot of $\log \phi$ against $1/T$ for various contraction levels we could obtain ΔE.

RESULTS AND DISCUSSION

Polypyromellitimide   The results for PI have been published in the previous paper(4). Figure 1 shows a typical tensile stress-strain curve for PI film. The measurement was carried out at 20°C with the very low rate of extension at $1/48000$ sec^{-1} After removal of the stress the strain recovers partially. However, the extended film shrinks markedly on heating. The sample is in a glassy state at this temperature range, then recoverable deformation should be energetic.

Then, it may be assumed that at the cold drawing the molecules gradually move from the equilibrium low-energy state to the metastable high-energy state at oriented situations, and are frozen in that state. On heating these frozen molecules would change back to the initial low-energy state and the film shows an overall shrinkage of the length. It was also found that the sample which was once heated to the high temperature shows no thermal shrinkage at the second heating.

The thermal shrinkage curves obtained for the cold-drawn specimens with different degree of stretching are shown in Figure 2. The ordinate shows the percentage contraction of the stretched films in reference to the original length before stretching. The stretched films show a shrinkage in the temperature range of 100 - 400°C.

The temperature derivative curves of thermal contraction were shown in Figure 3, obtained graphically from Figure 2. The shrinkage of the specimens stretched by 40% was measured at the various heating rates and are shown in Figure 4. From these curves, we obtain linear $\log \phi$ vs. $1/T$ plots, as seen in Figure 5, which give a range of the activation energy ΔE from

15 to 33 kcal/mol. Thermal shrinkage curves for 12.5% stretched samples gave ΔE of 10 kcal/mol and that for 30% stretched one gave a range of ΔE from 16 to 26 kcal/mol(4). We have concluded from these figures the three contraction processes appeared in 40% stretched sample in Figure 3 show the respective ΔE's of 10 25, and 33 kcal/mol. Wrasidlo(5) showed similar values of ΔE from his dielectric measurements of PI.

Polytrimellitamideimide Thermal shrinkage curves for stretched PAI films are shown in Figure 6. Figure 7 shows the temperature derivative curves, in which we find three contraction processes. Figure 8 shows curves for 40% stretched PAI films at various rates of heating. The kinetic plots for PAI in Figure 9 give ΔE's of 72, 17, and 13 kcal/mol. The process at the highest temperature range corresponds to the glass transition. The thermal shrinkage characteristics for PAI are similar to the corresponding data for PI. The effect of amide linkage is not clearly reflected to thermal properties.

Polybisphenol-A carbonate This polymer has p-phenylene linkages in its basic molecular structure. The film is not extensible as PI and PAI, and the 6.6% stretched film shows rather a thermal expansion, as seen in Figure 10. We reduced the thermal expansion term obtained for the as-received sample from the overall expansion curve for stretched samples to get a thermal shrinkage term. Figure 11 shows shrinkage curves obtained at different rates of heating. The temperature derivative curves in Figure 12 show two contraction processes. We analyzed these curves by the same way in Figure 13 and estimated ΔE of 17.4 kcal/mol for the high temperature process.

REFERENCE

1) H. Kambe and T. Kato, Applied Polymer Symposium, No.20, 365 (1973).
2) T. Ozawa, Bull. Chem. Soc. Japan, 38, 1881 (1965).
3) C. D. Doyle, J. Appl. Polymer Sci., 6, 639 (1962).
4) H. Kambe and T. Kato, J. Macromol. Sci.-Chem., A 8, 157 (1974).
5) W. Wrasidlo, J. Macromol. Sci.-Phys., B 6, 559 (1972).

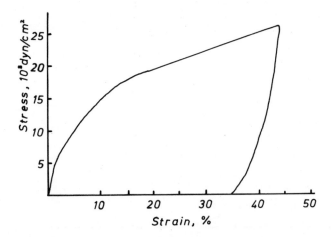

Fig. 1   Tensile stress-strain curve of PI.

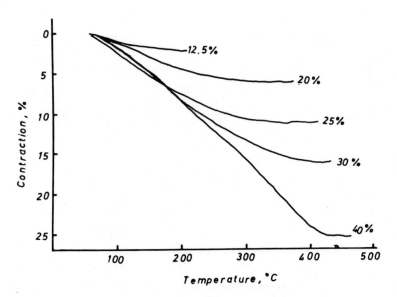

Fig. 2   Thermal shrinkage of PI films at various
degrees of stretching, 10°C/min.

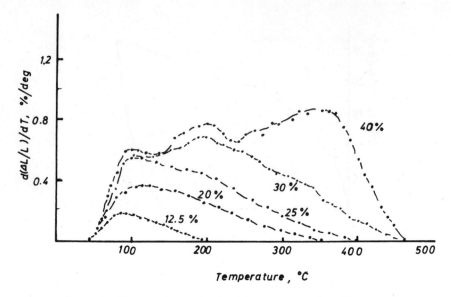

Fig. 3   Temperature derivative curves for PI.

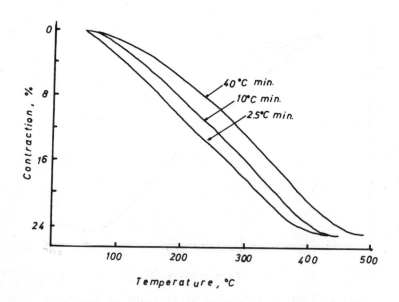

Fig. 4   Thermal shrinkage of 40% stretched PI film
at various rates of heating.

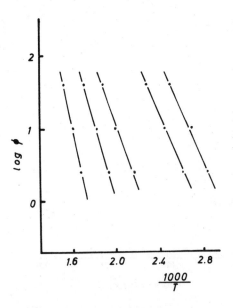

Fig. 5   Kinetic plots for thermomechanical analysis
of thermal shrinkage of PI.

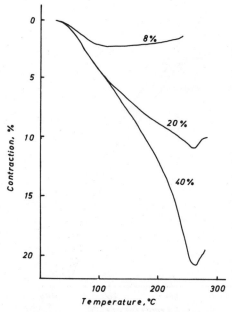

Fig. 6   Thermal shrinkage of PAI films at various
degrees of stretching, 10°C/min.

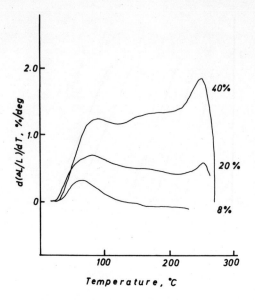

Fig. 7   Temperature derivative curves for PAI.

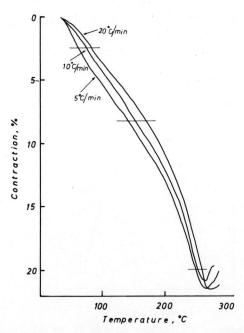

Fig. 8   Thermal shrinkage of 40% stretched PAi films
at various rates of heating.

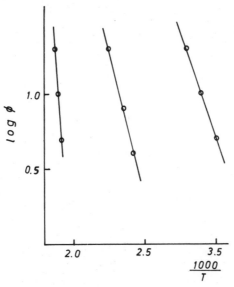

Fig. 9   Kinetic plots for thermomechanical analysis
of thermal shrinkage of PAI.

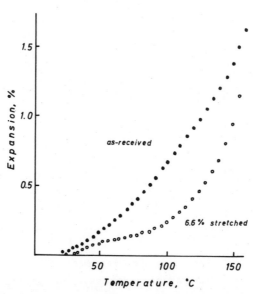

Fig. 10   Thermal expansion of PC as-received and 6.6%
stretched films.

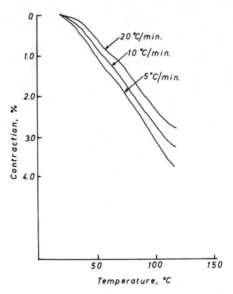

Fig. 11   Thermal shrinkage of 6.6% stretched PC films
at various rates of heating.

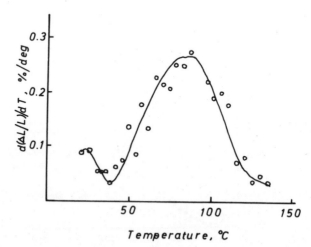

Fig. 12   Temperature derivative curves for PC.

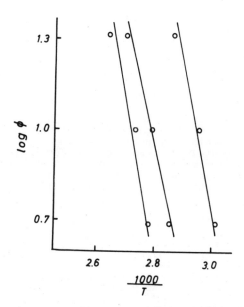

Fig. 13 Kinetic plots for thermomechanical analysis of thermal shrinkage of PC.

STUDYING THE EXPERIMENT

Paul D. Garn

The University of Akron
Akron, Ohio 44325, U. S. A.

ABSTRACT

Many of the conflicting observations in thermal analysis
arise from the manner in which the experiment was done.   The
high degree of reproducibility usually obtained by a single
observer is often in great contrast to the wide range of data
obtained by a typical group of laboratories, even for phase
transitions and meltings.   The variability in data from ICTA's
Second International Test Program shows the effects of differences in instrument design.

Differences in reported kinetic parameters may be due to
instrument design or neglect of atmosphere effects.   In descriptions of decompositions it is necessary also to examine the
mechanism of the reaction.   The probable sources of error in the
assumptions necessary to derive a kinetic equation are examined.
The decomposition of magnesium hydroxide illustrates the need to
consider the mechanisms prior to development of an equation.

THE VARIABILITY OF DATA

The data obtainable from a DTA, TG or other thermoanalytical
procedure are, in general, the resultant not only of the nature
of the process itself but also of the manner in which the data
are obtained.   In addition, the conclusions based upon the data
may be shaped or biased by the manner in which the data are
treated.

In this presentation, some effects of the details of the experiment will be examined.   This will include an extended
analysis of the published data of ICTA's Second International
Test Program.   The importance of consideration of the nature of
the process will be illustrated by discussion of the magnesium
hydroxide dehydration.   This has been the subject of a number
of investigations of its kinetics without regard for the special
nature of the process itself.   Some of the problems in kinetic
evaluations are considered.   The difficulty of describing the
shape by any small, general equation is demonstrated.

EXAMINATION OF THE CONDITIONS OF THE EXPERIMENT

The complexity of the DTA peak arises because:
1.   A temperature regime is customarily imposed upon a
point outside the sample.
2.   The heat from the furnace element must traverse

several materials and the interfaces between the element and the innermost part of the sample.

3. The several materials change their heat transport properties continuously but not identically with temperature.

4. The sample changes its heat transport properties discontinuously during a transition or decomposition.

5. In many apparatuses, the differential temperature is measured at a point which bears no unambiguous relationship to the sample itself. That is, the temperature of the measuring point has neither a fixed nor a smoothly-varying relationship to that of the sample.

6. Especially for decompositions, but even for phase transitions, the relationship between the temperatures of the sample and the measuring point changes because of the discontinuous demand for heat.

7. Unless the sample is enclosed in a homogeneous temperature environment, the loss of heat by radiation or conduction may cause dissymmetry in heat transport.

Some of these points may be illustrated by examination of the data from ICTA's Second International Test Program. The report by McAdie et al. (1) contains some interesting data which deserve some additional interpretation. The Committee on Standardization had confined its report (2) to the general relationships clearly discernible from the data. The more detailed report drew additional conclusions concerning the influence of a number of parameters upon the ONSET and PEAK temperatures on both heating and cooling, but did not explain the reasons for the observed behaviors. The purpose of this paper is to explain some of the variations in behavior between different apparatuses.

The data from 34 laboratories had been obtained on 35 instruments, four of which had been built in the user's laboratory. The greatest number of any one *model* of commercial apparatus was four, so the data were from a fair variety of instruments, the general similarities of the sample holders assemblies enabled classifications and comparisons. The comparisons have been by a single parameter at a time, so a pair of instruments having many features in common might, in some comparisons be in two separate classifications, each joined with instruments which may have only one feature in common. Consequently there are not enough data in many groups to justify firm conclusions *on the basis of these data alone.* From extensive related study by the present author, the additional conclusions given below are offered.

The Committee on Standardization reported to the ICTA the mean values of the ONSET and PEAK temperatures for the several Standard Reference Materials and some general observations on trends and reproducibility. McAdie et al. calculated the values shown in Table 1. From these data they concluded that the average standard deviations on heating for all compounds were ±6°C for the ONSET temperature, and ±7°C for the PEAK temperature, while on cooling the corresponding values were ±8 and ±10°C respectively.

TABLE 1

Mean Temperatures with Standard Deviations, Ranges of Reported Data, and Equilibrium Temperatures for the Standard Reference Materials. There are 57-71 Data Points for the Heating and 18-36 for the Cooling Cycles.

Compound	Heating		Cooling		T_{eq} [3]
	Onset	Peak	Onset	Peak	
KNO_3	128 ± 5 (112-149)	135 ± 6 (126-160)	122 ± 4 (112-128)	119 ± 4 (110-126)	127.7
In	154 ± 6 (140-162)	159 ± 6 (140-171)	154 ± 4 (146-163)	150 ± 4 (139-155)	157
Sn	230 ± 5 (217-240)	237 ± 6 (226-256)	203 ± 16 (168-222)	203 ± 17 (176-231)	231.9
$KClO_4$	299 ± 6 (280-310)	309 ± 8 (296-330)	287 ± 4 (278-296)	283 ± 5 (274-295)	299.5
Ag_2SO_4	424 ± 7 (400-439)	433 ± 7 (405-452)	399 ± 14 (337-413)	399 ± 15 (336-419)	430 [4]
SiO_2	571 ± 5 (552-581)	574 ± 5 (560-588)	572 ± 3 (565-577)	569 ± 4 (559-575)	573
K_2SO_4	582 ± 7 (560-598)	588 ± 6 (575-608)	582 ± 4 (572-587)	577 ± 8 (551-587)	583
K_2CrO_4	665 ± 7 (640-678)	673 ± 6 (656-692)	667 ± 5 (652-675)	661 ± 8 (630-671)	665
$BaCO_3$	808 ± 8 (783-834)	819 ± 8 (800-841)	767 ± 13 (742-790)	752 ± 16 (714-779)	810
$SrCO_3$	928 ± 7 (905-948)	938 ± 9 (910-961)	904 ± 15 (875-944)	897 ± 13 (868-920)	925

The principal reason for the somewhat greater deviations for the cooling data is the variation in the point of measurement. Nearly all the materials transform without measurable super heating, but several do supercool substantially. Tin, silver sulfate, barium carbonate and strontium carbonate supercool so badly that the good reproducibility of the cooling peaks of the other materials is hidden. As Table 1 indicates, the onset and peaks of the other materials are *more* reproducible than on heating.

The reason for the higher reproducibility is the nearer approach to temperature homogeneity within the sample holder assembly on cooling. The temperature distribution is distinctly different from the heating case because the entire assembly is the heat source, generating a heat flow by decreasing its temperature. This heat is radiated or conducted in all directions. On heating, however, this same radiation or conduction proceeds *except* for that part of the solid angle which the heater occupies. Note

92

that during cooling *less* heat may be dissipated to the heater than to the rest of the surroundings. Nevertheless, the nonuniformity of the temperature distribution is distinctly different from one type of sample holder to another.

But in steady state heating, there is at least some tendency toward temperature uniformity in the sample holder assembly, so all samples reach the reaction temperature at nearly the same indicated temperature. From this point on, the responses differ. The closer the measuring thermocouple to the sample itself, the less the indicated temperature interval of the reaction because the sample itself is influencing the temperature of the thermo- couple. This is expanded upon later.

Now consider the cooling. Again, the process will begin at nearly the same indicated temperatures. The exothermic process arrests the cooling and hence the indicated temperature if there is much contact at all between the sample and the thermocouple. Because of supercooling, the sample actually reheats, and this is detected in some arrangements, as indicated by the identity of the *average* ONSET and PEAK temperatures for both tin and silver sulfate. Radiation becomes so important at high temperatures that a lesser separation of the thermocouple from the sample is enough to cause the *measured* reheating to disappear. The statis- tics in the report (2) are (numerically) dominated by the well separated thermocouple. Whether in the reference or somewhere else in the furnace, these are not affected at all by the reheat- ing. They continue to register lower and lower temperatures, diminishing the average temperature and increasing the standard deviation of the whole set.

These conclusions are supported very clearly by Table 2, where the *interval* (defined *ad hoc* as the difference between the ONSET and PEAK temperatures) is tabulated for various locations of the temperature measuring thermocouple. The 00 location is typical of a sample block and deep cups with a recessed well. The 03 location is typical of shallow cups set on a thermocouple. Remembering that there are variations of sample holder type with- in these classes, the data still show that the central location 00 has the least change in the measured temperature on heating, but the greatest sensitivity to the supercooling. The sensitivity to supercooling implies clearly that the 00 position provides the greatest sensitivity to the actual temperature of the sample. The 03 position is clearly the next best. The others, in effect, simply follow the temperature program.

Concerning the spread of data, McAdie et al. concluded: The spread of temperatures reported for a given transition on heating varied from 22 to 51°C for the ONSET temperature and from 23 to 51°C for the PEAK temperatures. In both cases there was a tendency toward larger spread at higher transition temperatures.

The spread of temperatures reported for a given transition on cooling varied from 16 to 76°C for the ONSET temperature and

TABLE 2

Interval of Reaction by Compound and Location of Temperature-Measuring Thermocouple

HEATING

	00	03	10	21	30[#]	X[5]
KNO_3	5°C	8°C	10°C	5°C	7°C	3°C
In	3	4	7	5	4	---
Sn	3	7	10	7	5	---
$KClO_4$	4	9	15	7	8	1
Ag_2SO_4	6	9	12	6	8	5
SiO_2	2	4	3	3	2	3
K_2SO_4	4	6	11	4	5	3
K_2CrO_4	4	8	16	6	6	3
$BaCO_3$	8	9	12	11	12	6
$SrCO_3$	14	9	16	11	10	---

COOLING

	00	03[*]	10[#]	21	30[*]	X
KNO_3	2°C	0°C	4°C	3°C	4°C	0°C
In	4	1	4	4	---	---
SN	-16	-7	1	4	---	---
$KClO_4$	1	0	7	5	7	0
Ag_2SO_4	-9	-6	3	5	2	7
SiO_2	2	11	4	3	2	1
K_2SO_4	2	20	7	5	4	0
K_2CrO_4	3	23	7	6	4	1
$BaCO_3$	14	27	10	18	---	-2
$SrCO_3$	-13	14	9	9	---	---

00 In the sample, axially; 03 in contact with the sample, axially; 10 in the reference, axially; 21 location geometrically equivalent to sample and reference, non-axially; 30 location geometrically midway between sample and reference, axially. The data are calculated from the mean values without regard for the uncertainty as measured by the standard deviation.

* One observer
# Two observers (three for a few data points)
X Identified in text later.

from 16 to 83°C for the PEAK temperature. The freezing point of Sn and the $S_{II} \rightleftarrows S_I$ transitions in Ag_2SO_4, $BaCO_3$, and $SrCO_3$ exhibited particularly wide spreads in both ONSET and PEAK temperatures on cooling.

The substantial spread arises from a number of factors, of course, but it is important to note that some observers' data were consistently lower and others consistently higher than the means, as seen in Figure 1.

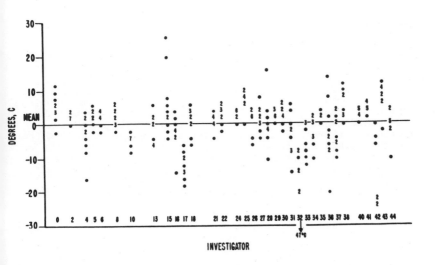

Figure 1. Deviations from the mean values by individual investigators. Each vertical array of points represents the several deviations of a single investigator's means for a compound from the mean of all the data for that compound. From McAdie et al.

The melting metals show the lowest spread. This is in part because some of the observers did not report because their apparatus would not contain liquids so that sample holders tended to be more alike. In general, Table 1 shows that the range for phase transitions, except for $BaCO_3$ and $SrCO_3$, is in the vicinity of 30-35°C, already high enough. The highs and lows may and usually do have good agreement between their data just as do those whose data fall near the mean. To illustrate, the present author's interval data on some of these compounds from the First

International Test Program are given in the last column of Table 2. The sample holder assembly was a block with an axial thermocouple, Type 00 in Table 2. The data are rounded to the nearest full degree. The data follow the same trends as do the total set.

The increasing spread of data with increasing temperatures may also be related to the different levels of sample-thermocouple interaction. The increase of the spread, though, is by no means proportional to the increasing temperature. The increasing importance of radiation helps to equalize temperatures within the sample holder assembly.

The spread of data for the cooling transitions of some materials has already been discussed, but it should be noted that, even with large supercooling, the transition interval may be small because of the operational definition of the ONSET. The interval may approach zero because the release of energy brings the sample to the PEAK temperature so rapidly that the measured ONSET is virtually coincident with the PEAK. The data for $BaCO_3$ in the last column of Table 1 illustrates this. The supercooling is great but irreproducible. The irreproducibility is undoubtedly related to the magnitude; that is, a material well below its transition point, yet at a temperature allowing rapid motion, can be expected to be especially sensitive to nucleation. In the seven runs, positive intervals as high as 9° and negative intervals as high as -23°C were found.

Concerning agreement with accepted thermodynamic equilibrium values:
With the exception of Ag_2SO_4, the mean ONSET temperatures on heating averaged -1°C of the accepted equilibrium transition temperature. For Ag_2SO_4 an equilibrium temperature of 430 ± 3 degrees appears to be more reliable than the 412 degree value cited in Reference 3.

With the exception of Ag_2SO_4, the mean deviation of the ONSET temperature on heating from the accepted equilibrium transition temperature was -1°C. Accepting that an equilibrium transition temperature for Ag_2SO_4 of 430 ± 3°C is more reliable than the 412°C cited by NBS, the mean deviation of the ONSET temperature from the equilibrium transition temperature was -1.5°C with a spread of -6 to +3°C. The mean deviation of PEAK temperatures on heating was +6°C with a spread of +1 to +9°C from the equilibrium transition temperature.

The good agreement of the DTA and thermodynamic values is partly fortuitous, as can be readily deduced from Table 2. The agreement should not lead any individual worker to dispute reported thermodynamic values on the basis of DTA data without thorough analysis of his instrument's characteristics and unless he can demonstrate very good agreement for other known values. Yet a preponderance of evidence, as with silver sulfate, should lead to reconsideration of the "accepted" value. Here again the

the domination of statistics by the separated thermocouple can be seen. In general, these measure the temperature of the sample holder. The phase change should begin when the holder first exceeds the equilibrium temperature. When the temperature within the sample is measured, the process is reaching completion at the measured temperature.

The data quite naturally have some dissymmetry. This is inevitable because real differences in conditions caused the changes as compared to statistical error. A measure of the dissymmetry can be obtained from the spread shown in Table 1, where, in general, the lowest temperature is half again as far from the mean as is the high temperature. A greater number of data points in the higher regions has brought the mean upward. From the other point of view, the greater amount of close-lying data yielding higher temperatures has been influenced by some outlying lower temperatures. There are over half again as many data points above the mean in Figure 1 as below.

It is reasonable and proper to conclude that variations in the ONSET and PEAK temperatures found in the ICTA's Second International Test Program are related to the degree of separation of the measuring thermocouple from the sample. The cooling mode yields data at least equal in precision to the heating mode except for materials which super-cool greatly.

EXAMINATION OF THE PROCESS

Background

A major peril in formulating descriptions of reactions or transitions is the expectation that all examples of either class will fit a single pattern. The lack of similarity of even $Solid_1 \rightleftarrows Solid_2$ transitions has already been pointed out (1). That is, some transitions are quite commonly limited by the heat supply under ordinary conditions, some superheat and many supercool, and some show behaviors which are somewhat time dependent either in the progress of the transition or as a function of the time the material has been in the initial state.

In chemical reactions the lack of similarity is still more pronounced. As compared to phase transitions the heat requirement (for endothermic processes) is ordinarily much higher; decompositions may require 5-50 kcal/mole as compared to a few hundreds of calories per mole for transitions. Some show very decided chemical rate dependence while others show behaviors limited principally by the experimental conditions. Others show a rate dependence not upon the chemical process but upon the supply or removal of reactants or products.

Because of this variability in types of processes, it is essential that the nature of the rate-controlling process be discovered rather than assumed. That is, we should be able to work toward descriptions of each of the several types of transitions or reactions, but the relationship is obscured if we try to fit all transitions or all metathetical reactions or all decomposition reactions into single patterns for each class. General descriptions of the smaller groups based upon the type of rate control may be achievable.

Thermal decompositions are studied both from theoretical interest in the chemistry and mechanisms of reactions and from a practical interest in carrying out the reactions to gain the product. For these reasons, there have been many attempts to gain useful information on the kinetics of decomposition processes. These attempts have taken, in general, two directions -- one involved very specifically with the mechanism of the process, the other involved with the overall kinetics of decomposition processes.

This latter approach often makes use of classical concepts from homogeneous kinetics -- an order of reaction and the Arrhenius equation. The reality of an order of reaction is totally inconsistent with a reaction interface or any localized reaction, although a few special conditions of reaction can be described by the mathematics of certain "orders of reaction." Basically, however, the instantaneous rate of a solid state reaction is not dependent upon the amount of unreacted material in the whole of the sample, but upon conditions at the reaction site.

Kinetic Order

Kinetic equations which are based upon orders of reaction are incorrect; hence the quantities calculated from experimental data will vary with experimental conditions. For example, the apparent activation energy has been found to be dependent upon heating rate and even upon sample size. The bases for some of these errors have already been described (6).

First order kinetics could describe any reaction in which nucleation of the new species is very slow in comparison to the growth of the nuclei. Two-thirds-order and half-order kinetics could describe uniformly advancing interfaces in spheres and cylinders respectively. Very often, however, the escape of a product or the supply of a reactant or the transfer of energy limits the reaction rather than the chemical process.

If the process is truly chemical-reaction-rate limited, the kinetic equation should not depend upon atmosphere or sample shape or size or manner of holding it. Hence the test for applicability of a kinetic equation is simple. One can increase the sample size by an order of magnitude with confidence that the rate equation will still be descriptive. One can use vacuum or a pressure of the gaseous reactant or product or change the sample holder from a free-diffusion to a self-generated atmosphere type without affecting the calculated kinetic parameters. But if one of these tests fails, the description of the process is inaccurate.

Diffusion and Nucleation-and-Growth

In a greater proportion of kinetics studies, the homogeneous reaction concept is discarded in favor of reaction interfaces whose advance may be limited by diffusion of reactants or products, by a real chemical reaction rate, or by inception (nucleation) of the interface. A variety of approaches which have been used in an attempt to describe the course of the reaction in terms of the quantity of unreacted material.

In each of these, some assumed model is used to derive the equation, hence the validity of the equation is dependent upon the degree of exactness of at least the principal assumptions. Note that even good agreement of the final equation with a set of data is not sufficient proof of its validity. As a general truth, more than one equation can be found to fit a particular set of data fairly well. Guarini and Spinnicci (7) have illustrated the difficulty of ascertaining which of several proposed equations fit their data best. Brindley etal. (8) have found that their data on kaolinite did not fit diffusion or first-order reactions well, although the agreement with either two- or three- minemsional diffusion was fairly good. Anthony and Garn (9) have reported that a nucleation-and-growth model fitted the kaolinite dehydroxylation (under controlled water vapor pressures) better than other extant models, but they pointed out that "It is unlikely that a single mathematical description of this dehydroxylation can be formulated." Selvaratnam and Garn (10) have shown that even under controlled water vapor pressures similar to those used by Anthony and Garn the course of the reaction was influenced by the geometry of the specimen, presumably because the dehydroxylation was limited by the heat supply.

In considering the several models, one must pause to examine the assumptions implicit in the model as well as those explicitly stated by the writer. In addition, any stated or implied approximations must be evaluated. For each evaluation, some thought concerning the magnitude of the error is needed. Small uncertainties may be overlooked; there is no real hope of exact treatments because the system is continually changing. Nevertheless, a quite unwarranted assumption is a reason to suspect the result, and an assumption which is clearly contrary to the real condition is sufficient reason to discard the treatment altogether.

It is therefore instructive to examine a selected equation to ascertain to what extent the mathematics can be trusted to describe a real system. There are several which have been developed for reaction modelling, but have found their way into the thermal decomposition literature. These equations are often treated empirically, that is, a set of data is tested against equations until one is judged to fit better than the others. Consider, as our chosen example, the Ginstling-Brounshtein equation (11)

$$1 - \frac{2}{3}\chi - (1 - \chi)^{2/3} = kt$$

with the understanding that objections raised to this equation may apply equally well to any similar model.

The Ginstling-Brounshtein Equation

The Ginstling-Brounshtein equation was developed for an
A + B → C reaction rather than for thermal decompositions. The
model comprised spherical particles of uniform size and that the
reaction occurred at a discrete interface. The whole list of
conditions·is given below with discussions of the merits and
deficiencies of each. Even with the several limitations which
simplify the model upon which the final equation is based, it has
been used frequently to describe thermal decomposition processes.
Prima facie justification can be found by treating the reverse
reaction and considering the gaseous product as the diffusing
species. On that assumption, we can examine the postulates:

1. The reaction under consideration can be classified as
an additive reaction.
This is already treated above. We are now dealing with a
subtraction reaction.
2. Nucleation, followed by surface diffusion, occurs at a
temperature below that needed for bulk diffusion. A coherent
product layer is present when bulk diffusion does occur.
The nucleation process is apparently quite easy for most
decompositions. Surface imperfections, incomplete co-ordination
etc., make it possible for some solid product to be present
even when the reaction is not proceeding at a measurable rate.
Completion of a surface layer of product would consume a very
small fraction of reactant.
3. The chemical reaction at the phase boundary is considerably faster than the transport process, and thus the solid-state
reaction is bulk-diffusion controlled.
This sets up the condition that we are not studying the
chemical reaction rate at all. The limiting step in reactions
described by this equation is diffusion of reactant (or, in
decompositions, gaseous product) through the product layer.
This further means that any apparent activation energy calculated is that for diffusion, rather than for chemical reaction.
4. Bulk diffusion is unidirectional.
This is valid with some qualifications.' The anisotropy
of crystals suggests preferred paths, so the diffusion, though
always outward, is not uniformly distributed.
5. The product is not miscible with any of the reactants.
This is generally acceptable.
6. The reacting particles are all spheres of uniform radii.
This condition will introduce small error in the early stages
if the particle is not quite round but is isotropic. After the
early reaction, the error will be diminished and possibly
negligible if the product is isotropic or forms many crystallites so that gaseous diffusion takes place between, rather than
through, the product crystals.
7. The ratio of the volume of the product layer to the volume
of the materials reacted is unity.
The postulate of unchanged volume is hardly ever acceptable.
In some cases the change may be small and in other cases the
lattice of the reactant may be retained. In most cases there
is a substantial volume change.

8. The increase in the thickness of the product layer follows the parabolic rate law.
This behavior is implicit in the bulk-diffusion-controlled model. The increase in thickness need not be uniform in all directions, but along a given direction the parabolic rate law should be descriptive.
9. The diffusion coefficient of the species being transported is not a function of time.
A constant diffusion coefficient will be descriptive of some cases but not all. The diffusion coefficient is necessarily related to the condition of the product layer. For those materials which go immediately into a well-formed crystalline product, whether finely or coarsely crystalline, the condition is acceptable. For many reactions, however, there is evidence for annealing of crystallites removed from the reaction interface. For other reactions, strong adsorption of the gaseous product is postulated. In either case, there will be some time dependence for the diffusion coefficient for that system.
10. The activity of the reactants remains constant on both sides of the reaction interface.
The condition of constant reactivity should be descriptive of most reactions as soon as a quasi-steady state has been set up. It must be kept in mind that this condition is limited to the interface; the annealing, recrystallization, etc., which may take place after the interface moves on will have effects already described.

It is clear that a reaction which could be described precisely by the Ginstling-Brounshtein equation would be a rarity indeed, either for the forward or the reverse reaction. But accepting for the moment that it is the best description of some forward reactions, let us examine the difference at the interface for the forward and reverse cases. In the forward reaction, the diffusing reactant is continually in a state of depletion because the chemical reaction is rapid. The rate of chemical reaction is not temperature dependent under the conditions of the experiment. The observed rate dependence is that due to diffusion. Diffusion is a "activated" process, that is, it becomes more rapid with increasing temperature. In this it resembles homogeneous kinetics. The temperature dependence, however, is related to the activation energy for diffusion and hence to the experimental conditions related to diffusion. No conclusion regarding the activation energy of the chemical process should be drawn.

Now examine the reverse process where the diffusing species is a gaseous product. The gaseous product is in good supply at the reaction interface but this supply is not likely to be constant in concentration. The reaction and the diffusion must set to a quasi-steady state to match supply (by reaction) and withdrawal (by diffusion). For a diffusion limit to be applicable, it is essential that some step in the decomposition process be reversible otherwise the unrestricted increase in gaseous product would lead to continued pressure increase and eventual rupture of the crystal or an enlargement of the diffusion paths. But this would be a time-dependent change in the diffusion coefficient. See condition 9.

101

Some reactions are rapid and reversible enough to be describable by an equilibrium constant under most conditions, as compared to having some reversible step which may limit the overall process. In the equilibrium case we have two influences which lead to a temperature dependence of the observed rate and hence to the opportunity to calculate a quantity which can be called the activation energy. The increase in temperature not only "activates" the diffusion but also increases the pressure of the gaseous product, so the rate of escape (hence the measured rate of reaction) is enhanced. Since the two effects cannot be separated, the observed temperature dependence has no significance beyond that of an empirical temperature dependence.

Careful examination of the models used in other derivations of kinetic equations will disclose similar deficiencies, but in addition to the difficulty of describing the process of escape of the gaseous product it is also not yet possible to provide a general model for the initial formation of the gaseous product. One case in which the mechanism must be taken into account is the dehydration of magnesium hydroxide.

Dehydration of Magnesium Hydroxide
(Background)

The dehydration $Mg(OH)_2$ has been studied frequently. The number of conclusions concerning the kinetics is somewhat less than the number of reports. Gregg and Razouk (12), for example, found that their data could be described by a contracting sphere equation while Anderson and Horlock (13) fitted their data to a contracting disc equation. But Komatsu (14) reported that the kinetics could be described equally well by the contracting disc, the contracting sphere or the first order decay models. Sharp and Brindley (15) calculated similar activation energies for 1, 2/3, and 1/2 order reaction, and the parabolic diffusion law, finding a range of 20-23 kcal/mole for these models.

There is a water vapor pressure effect to be concerned with, too. This author (16) has previously discussed pressure effects in equilibrium or rapidly reversible systems. At low pressures the decomposition proceeds at temperatures ca 300°C. The system is not rapidly reversible, although rehydration does take place at the lower temperatures. The equilibrium temperatures at low pressures have not been reported.

At high pressures, the equilibrium temperatures are greater than 500°C. (17-19) This refers to the bulk system. It appears that the dehydration actually occurs below the equilibrium temperature for the bulk material. Any description of the process should take into account this difference in temperature.

Freund (20), Freund and Nägerl (21), and Nägerl and Freund (2 have postulated a tunnelling mechanism for the $Mg(OH)_2$ dehydration Based upon spectroscopic evidence, the energy difference between the zero energy level and the lowest excited levels suggests populations of 10^{-2}, 10^{-4}, and 10^{-6} for the first, second and third excited levels, respectively, and even these levels do not allow mobility in the classical sense. The observed rates of decomposition, hence, are unlikely to be due to the mobility of the proton throught the lattice.

On the other hand, quantum mechanics allows the possibility of movement (of elementary particles) between locations of equivalent energy in spite of the energy barrier to classical movement. Corresponding energy levels must exist, one of which is empty. The energy levels need not have the same energy at all temperatures. Different temperature dependences, band broadening and combination (coupled) modes of vibration can all contribute to the incursion of band overlap at some elevated temperature.

In the crystal lattice, the OH^- groups are equivalent crystallographically. The lower energy levels are equivalent but filled. The energy barrier to movement must be near the ionization energy (of the OH^-) and the population of excited levels is low. Physically, the amplitude of oscillation of particles, stretching or bending, is restricted by the fields of the neighboring OH^- groups (Figure 2). On the surface, this restriction is removed.

Again considering Figure 2, but with the top Mg and OH layers removed at the cleavage plane, it is clear that the stretching mode at the surface is no longer impeded, but it is only a little less clear that the bending mode is also less restricted (Figure 3). Considering now the bending of the O^{2-} with respect to the Mg, with the understanding that the small proton is moving within the electron cloud of the O^{2-}, maintaining its wagging and stretching vibrations, the OH^- may vibrate in two orthogonal modes, the in-phase bending which results in no change in OH^- - OH^- distance and an out-of-phase vibration which results in repeated approach and retreat of adjacent OH^- groups. In two dimensions, these vibrations must inevitably couple, and couple to a degree not experienced within the crystal because of the restricted movement of each group.

Within the crystal, the frequency limits for these in-phase and out-of-phase vibrations are 368 cm and 575 cm^{-1}, respectively, according to Nägerl and Freund (22), who, from the force constants and reduced mass, estimated the central frequency of the coupled bending as 480 cm^{-1}.

If the decomposition takes place by proton transfer, as is reasonable from the evidence, we may postulate that some amplitude of vibration is required to enable to proton transfer. For proton tunnelling, this would be at a temperature which, with band broadening, yields an overlap of a filled and an unfilled lever. The mean square amplitude u^2 of a vibration can be calculated from James' formula. Figure 4 shows Nägerl and Freund's plot for the upper and lower limiting frequencies and for the combination frequency, 480 cm^{-1}, calculated for the comparatively unrestricted surface vibrations.

103

Remembering that the equilibrium temperature for the dehydration is above 500°C, it is reasonable to select a temperature just within that range and assume that the calculated amplitude for the out-of-phase vibration in sufficient to provide the overlap of donor and acceptor levels needed for proton tunnelling. (It is obvious that the in-phase vibration already exceeds this amplitude at room temperature, but in this mode the OH^- - OH^- distance remains constant so the tunnelling probability is vanishingly small.) It is reasonable to assume that the combination frequency possible at the surface need only reach that same amplitude (Figure 4) before corresponding (overlapping) donor-acceptor energy levels are found. If 800°K (527°C) is taken as the dehydration temperature within the crystal, it is clear that the combination frequency reaches this temperature below 600°K or ca. 300°C. This is in accord with experimental evidence.

Recently, Girgis (23) concluded from DTA evidence that none of the methods which depend upon an order of reaction was satisfactory for describing his data. He calculated activation energies much greater than the enthalpy change and also higher than values reported elsewhere (12,13,24). Sharp (26), in reply, summarized the activation energies reported from isothermal weight loss measurements and noted that these were in the 20-30 kcal/mole range, much lower than those reported by Girgis. He concluded that the methods used to obtain kinetic parameters from DTA should be applied with great caution and only under favorable circumstances.

The computations reported by Girgis are seriously affected by the experimental conditions. The apparatus used, like other small-sample quantitative DTAs and DSCs, sets up a temperature gradient in the apparatus and measures the temperature difference between symmetrically placed sample and reference locations. The heat consumption by the DTA or DSC specimen being a small part of the total energy, it becomes a perturbation of the temperature distribution within the apparatus. The temperature gradient is characteristic of the heating rate, the temperature, and the facilities for dissipation of heat. Bohon (26) has shown that changes of gases in this apparatus cause significant changes in the calibration constant and Dosch (27) has shown similar behavior for another quantitative DTA. In each case, a large fraction of the thermal energy is dissipated through the gas to the cooler parts of the assembly. Hence, even for a very rapid transition or melting, a heating rate dependence is observed because the sample is in effect, off to one side of the heat path, while the temperature is being measured at a point closer to the main heat flow (28). This behavior is a common problem in small-sample DTA or DSC measurement.

Girgis' curves at heating rates of 5, 10, and 20°C/min. (Figure 5) show an increase in both the extrapolated onset and peak temperatures. The data suggest that the peak temperature increases by about 25°C when the heating rate is increased four-fold. Part of this increase is ascribable to the sample; the reversibility of the reaction suggests a van't Hoff dependence on water vapor pressure at the reaction site, so the increasing pressure of water vapor tends to retard formation of more vapor, and the very real time-dependence of the reaction, which appears as an apparent increase in temperature because the measuring point reflects the influence of the whole sample holder assembly rather than only the temperature of the sample. Some separation of these effects is possible by direct control of the water vapor pressure and by measuring directly the temperature of the sample.

Maintaining a pure water vapor pressure provides a constant environment for each particle; changing the pressure therefore exercises a direct control over the overall process within the particle, enabling a clear determination of the need to include water vapor pressure in the rate equation for the decomposition. The reasons are discussed below.

Attempts to derive meaningful kinetic data for reversible processes are generally based on experiments in which the reverse reaction is assumed to be negligible because the products of the forward reaction are swept or pumped away from the exterior of the sample. The procedure has limited validity because gaseous reaction products will tend to remain within the (powdered) specimen or within the individual particle. As the reaction proceeds, the product gases move away from the reaction site in part by diffusion but in part also by reason of an excess pressure within the particle. The pressure gradient between the reaction site and the exterior of the sample is not directly measurable, so any estimates must be based upon assumptions which cannot be expected to have general validity. For example, the pore size developed in kaolinite is unlikely to be close to that developed in brucite -- also by water vapor. Likewise, there is no a priori reason to believe that the pore size or configuration developed in a salt hydrate, $CaC_2O_4 \cdot H_2O$, for example, is closely similar to the kaolinite or brucite cases.

Such experiments will have a known maximum pressure external to the sample, rather than a known pressure at the reaction site or interface. In high vacuum experiments, the unknown pressure difference is virtually the total pressure at the reaction site, but a similar pressure difference at higher total pressures of the product gas is a smaller part of the whole pressure. This unknown pressure difference can become the lesser quantity and, at higher pressures, can become negligible. It is reasonable, therefore, to study decomposition reactions at pressures near atmospheric as well as in vacuum.

In this work the pressure dependence of the $Mg(OH)_2$ dehydration was first studied at 1, 2, 4, and 8 atmospheres, that is, with equal increments of log P, providing a direct test of the applicability of the van't Hoff equation.

(Experimental)

Magnesium hydroxide powder, Merck 5870, was used in the principal experiments. The single crystal is from Texas, Lancaster County, Pennsylvania, USA, Smithsonian Cat. No. 14390,

supplied by courtesy of P. Desautel. For powdered samples,
56   1 mg of Mg(OH)$_2$ were packed with light pressure into the
sleeve.

The principle behind the maintenance of the water vapor
pressure is similar to that already described (29) for a dynamic
atmosphere (Figure 6). A dynamic atmosphere is not needed here
because the reaction under study involves a single gas; the furnace assembly therefore has only a single outlet for water vapor.

The furnace assembly comprises a ceramic furnace tube sealed
to a platform secured to the mounting panel, a sample support
assembly secured to the furnace tube platform, a water reservoir
with a level gauge, a pressure source (nitrogen cylinder with
regulator) and appropriate valving. The pressure at the furnace
is monitored by a second gauge connected to the supply tube
between the furnace valve and the furnace. The thermocouples are
Chromel-Alumel, shielded, 0.50 mm. o.d., sealed into the support
tubes protruding above the end of the tube ca. 1.0 cm. to enable
measurements in a block as well as in cups or other devices. The
support tubes are machined to enable positioning of a multihole
splash guard which fits closely inside the ceramic tube so that
it also positions the thermocouples. The temperature is also
equalized by an aluminum ring ca. 30 mm. high, 18 mm. o.d., and
3 mm. thick. The exit tube extends to near the top of the furnace
chamber. Flow is restricted at the top (interior) end of the
tube. A more detailed description is given by Garn and
Freund (30).

In this work, powdered samples were held in stainless steel
sleeves ca. 3 mm. i.d., closed (but not sealed) by aluminum plugs.
The thermocouple passed through one of these plugs so that the
junction was axial and approximately at the (vertical) center.
The sample tubes with plugs also enable an approximation to
self-generated atmosphere (31) operation. In this use, nitrogen
is supplied directly to the furnace chamber, repressing the
escape of the water vapor from the sample until its vapor pressure
exceeds the applied pressure.

When single crystals were tested, they were mounted directly
upon the thermocouples by boring a hole, close-fitting on the
thermocouple, through the crystal and providing vertical support
with a ceramic insulating ring encircling the thermocouple.

In preliminary experiments, high purity nitrogen was supplied
to the furnace chamber at ambient pressure and at 7 kg./cm^2 gauge
(8 atm. total). Open cups were used to enable free escape of
water vapor and closed sleeves to approximate a self-generated
atmosphere of water vapor (31) The confinement inhibits the
rapid escape of water vapor until the pressure in the sleeve
exceeds the furnace chamber pressure.

For water vapor studies, the system was closed and a nitrogen
pressure is introduces at the top of the reservoir and the pro-
grammer advanced to about 130°C. When a steady state heating

had been reached, the pressure regulator was adjusted to maintain the selected pressure at the furnace gauge.

(Results)

The importance of atmosphere control can be easily shown from self-generated atmosphere operations. In preliminary experiments, the commercial magnesium hydroxide was heated at atmospheric pressure and with 7.0 kg/cm^2 applied pressure (about 8 atm.) both in open cups and the closed sleeves. The increased pressure increased the peak temperature in the open cups by 10°C (from 421 to 431°C) showing that the reduced diffusion out of the particle influenced the reaction. Operation in the closed chamber increased the decomposition temperature by 20°C at one atm. and by 40°C at 8 atm. This shows that the atmosphere surrounding the particle also affects the reaction within the particle. That is, the temperature of dehydration of Mg(OH)$_2$ is clearly dependent upon the atmosphere surrounding the particles. Hence for any thermodynamic or kinetic study of the process, the influence of the atmosphere must be known. This study shows some aspects of that influence and shows that inadequate control yields inferior data.

The forms of the DTA and the temperature curves of Figure 7 indicate that the dehydration proceeds at a not-quite-constant temperature. It is also clear that the temperature of the specimen is not rising at the programmed heating rate of 20°/min. During the 2 1/2 min. from onset to peak, the temperature at the center has risen only 13°. The obvious conclusion is that any kinetic treatment assuming T = a + bt, where a is the initial temperature, b is the programmed heat rate and t is the time, does not describe this process; heat or mass transport limits the reaction.

The heat transport can be eliminated or confirmed as a major effect by ascertaining the behavior of a diluted sample. One-third the amount of Mg(OH)$_2$ required to fill the sample holder was mixed with twice its volume of alumina. The resulting curve showed very nearly the same behavior as the undiluted sample. The relatively small influence of heat transport limiting is also indicated by the moderate change of peak temperature with heating rate. At 2 atm., an eight-fold change raised the peak temperature 13°C, the values being 416, 420, 425, and 429° at 2.5, 5, 10, and 20°C/min., respectively. Heat transfer controlled processes show little change in axial temperature with heating rate.

Calculation of the activation energy, by the method of Kissinger (32) yielded a value about twice the maximum value found by Girgis (23) by the same method and hence far above those generally found by isothermal dehydration in vacuo. This is typical behavior; for some decompositions the rate of reaction is virtually zero until the temperature is high enough to cause some primary process to occur (33) but then the rate of reaction does not change with temperature in a manner which can be described by the Arrhenius equation without introducing a phenomenon called the "kinetic compensation effect" in which the frequency factor is allowed to vary over a wide range to adjust the temperature dependence (calculated activation energy) to the realities of the experiment. Another effect which leads to high calculated activation energies is presence of the gaseous decomposition product either as a controlled or self-generated atmosphere or by experimental conditions which inadvertently favor the accumulation of the gaseous product. For example, Anthony and Garn (9) calculated apparent activation energies of 60 - 260 for kaolinite (isothermally) with changes in water vapor pressure from 0.4 to 3.2 atm.

The pressures first used were such that $\Delta \log P$ was constant, but the first set of data suggested a more nearly linear variation with pressure. Subsequent data sets included runs at 6 atm. (5.0 kg/cm^2 gauge). The peak temperatures listed in Table 1 are plotted in Figure 8 vs P and vs $\log P$. The curvature of the $\log P$ plot shows clearly that the van't Hoff relationship is not descriptive. The nearly linear variation with pressure is also consistent with a relatively small contribution from the reverse reaction. The system is far from equilibrium.

The similarities of the peaks are obvious from Figure 9, in which the curves from a single set are shown. They are plotted against time and only a single index mark (400°C) given because the variation in temperature is slightly different in each case.

One phenomenon which must be considered is the remarkable constancy of total reaction time. As an arbitrary test, a point near the end of the peak corresponding in definition to the extrapolated onset at the beginning of the peak was measured for the peaks shown in Figure 9. At a 5°/min. heating rate, this reaction interval between the extrapolated onset and the extrapolated end was 8 1/2 to 10 min. at all pressures. Four single crystals showed reaction intervals of 9-10 min. at temperatures ca. 40° higher than the powder. This suggests a reaction which above some lower limit in temperature, reaches a velocity which is almost independent of temperature. This result supports a reaction mechanism involving proton tunnelling as the initial step (20-23). A behavior with temperature and pressure as found in this work could be caused by tunnelling between a very narrow and a broader band of energies. The measurable overlap is reached at some temperature T_t, but at increasing temperatures complete oveverlap with the narrower energy band is quickly reached. The observed rate then becomes the temperature dependent diffusion of water vapor from the reaction site. This is consistent with the uniformity of behavior of single crystals of various sizes.

The extrapolation of Figure 8 to very low pressures is unrealistic; decomposition as low as 300°C in vacuo is known to occur. The pressure at the reaction interface is not known, but it is noteworthy that brucite samples weighing 51 mg and less decomposed rapidly enough at 300°C to raise the pressure in a pumped system (8) by an order of magnitude. At higher temperatures the pressure increase was 200-fold.

Nonetheless, the pressure dependence apparently changes,

probably by a changing rate-limiting process. That is, at lower temperatures the rate limit may well be the primary step, the formation of water vapor at the surface of the unreacted $Mg(OH)_2$. The topitactic nature of this reaction has been amply demonstrated (13,34,35).

It is reasonable to infer that a changing process allows a relatively great change of rate with temperature at the lower temperatures (and pressures) but a smaller change at the higher pressures (and temperatures). The variation in process is much more likely to result from a new limiting step (diffusion) than from any discontinuous change of reaction mechanism. The curvature of the $\underline{T}$ $\underline{vs}$ $\underline{P}$ plot in Figure 8 should also be noted. The apparent tendency to level off at higher pressures is being studied further.

We may conclude from these data that:

The dehydration of magnesium hydroxide is limited in its observed rate by diffusion of water vapor when that water vapor pressure is in the vicinity of the ambient pressure. The increasing water vapor pressure at the reaction site may tend to slow the reaction by adsorption on the MgO sites adjacent to the still unreacted $Mg(OH)_2$. The molecule would tend to limit the amplitude of the OH^- wagging vibration.

Extrapolation toward lower temperatures does not connect with data obtained in high vacuum so a changing of the rate limiting process is inferred. The tunnelling process which requires excitation to a higher energy level at lower temperatures may become rapid enough at higher temperatures that diffusion of the water vapor against the imposed pressure becomes rate limiting.

The decomposition of $Mg(OH)_2$ is unlikely to be describable by any extant kinetic equation except fortuitously and certainly not by any based upon a calculated or assumed order of reaction. The temperature dependence from which apparent activation energies have been calculated is related instead to the increasing back pressure at increasing temperatures.

REFERENCES

1. H. G. McAdie, P. D. Garn, and O. Menis, United States National Bureau of Standards Special Publication 260-40. U. S. Government Printing Office, 1972
2. H. G. McAdie, Volume 1, page 591, Thermal Analysis. Birkhäuser Verlag, Basel and Stuttgart, 1972. Edited by H. G. Weidemann.
3. U. S. National Bureau of Standards Circular 500, Selective Values of Chemical Thermodynamic Properties. U. S. Government Printing Office, 1952
4. J. A. Hedvall, R. Lindner, and N. Hartler, Acta Chem. Scand. 4, 1099 (1950)

5. P. D. Garn, Technical Report AFML-TR-68.139. Air Force
    Materials Laboratory, Wright-Patterson Air Force Base,
    Ohio, U. S. A.
6. P. D. Garn, CRC Critical Reviews in Analytical Chemistry,
    65-111, September 1972
7. G. G. T. Guarini, R. Spinnicci, F. M. Carlini and D. Donati,
    J. Thermal Anal. 5, 307-314 (1973)
8. G. W. Brindley, J. H. Sharp, J. H. Patterson and B. N.
    Narahiti, Am. Min. 52, 207-277 (1967)
9. G. D. Anthony and P. D. Garn, Kinetics of Kaolinite Dehy-
    droxylation, J. Amer. Ceram. Soc., 57, 132-135 (1974)
10. M. Selvaratnam and P. D. Garn, Kinetics of the Kaolinite
    dehydroxylation II., to be published.
11. A. M. Ginstling, B. I. Brounshtein, Zh. Prikl. Kim,
    Leningrad, 23, 1249-1259 (1950)
12. S. J. Gregg and R. I. Razouk, J. Chem. Soc. (London) 1949,
    S36-44.
13. P. J. Anderson and R. F. Horlock, Trans. Faraday Soc. 58,
    1993-2004 (1962)
14. W. Komatsu, Private communication cited by Gordon and Kingery
    Ref. 32.
15. J. H. Sharp and G. W. Brindley, Sci. Tech. Aerospace Rept. 4,
    706 (1966)
16. P. D. Garn, Thermoanalytical Methods of Investigation, Chap.
    7, Academic Press, New York, 1965
17. H. L. Barnes, and W. G. Ernst, Am. Jour. Sci. 261, 129 (1963)
18. W. S. Fyfe, Am. Jour. Sci. 256, 729 (1958)
19. D. M. Roy, and Roy, Rustum, Am. Jour. Sci. 255, 573 (1957)
20. F. Freund, Proc. Int. Clay Conf., Tokyo, 1969, Vol.I, p. 121
21. F. Freund, and H. Nägerl, in "Thermal Analysis" (Robert F.
    Schwenker and Paul D. Garn, eds.), Vol. 2, p. 1207, Academic
    Press, New York, 1969
22. H. Nägerl and F. Freund, J. Thermal Anal. 2, 387 (1970)
23. B. S. Girgis, Tans. J. Brit. Ceram. Soc., 71, 177 (1972)
24. E. Kay and N. W. Gregory, J. Phys. Chem., 62, 1079 (1958)
25. J. H. Sharp, Trans. J. Brit. Ceram. Soc., 72, 21-23 (1973)
26. R. L. Bohon, Proc. 3rd. Toronto Symposium on Thermal Analysis
    33 (1969). Edited by H. G. McAdie
27. E. L. Dosch, Thermochim. Acta, 1, 367-371 (1970)
28. P. D. Garn, Calorimetry, Thermometry and Thermal Analysis,
    Japan Society for Calorimetry and Thermal Analysis 1971,
    pp 45-56
29. P. D. Garn, Rev. Sci. Instr., 44, 231-233 (1973)
30. P. D. Garn and F. Freund, Trans. J. Brit. Ceram. Soc., (in
    press, 1974)
31. P. D. Garn and J. E. Kessler, Anal. Chem., 32, 1563 (1960)
32. H. E. Kissinger, Anal. Chem., 29, 1702 (1957)
33. P. D. Garn, On the Kinetic Compensation Effect, J. Thermal
    Analysis, in press, 1974
34. R. S. Gordon and W. D. Kingery, J. Am. Ceram. Soc. 49,
    654-660 (1966)
35. R. S. Gordon and W. D. Kingery, J. Am. Ceram. Soc., 50,
    8-14 (1967)

ACKNOWLEDGMENT

The author is grateful to the U. S. National Science Foundation and to Germany's Alexander von Humboldt Foundation for support of parts of this work and to the Journal of Thermal Analysis for permission to use substantial parts of a publication in press.

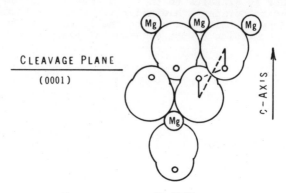

CLEAVAGE PLANE

(0001)

c, – AXIS

MODEL OF THE Mg(OH)$_2$ LATTICE.

Fig.2 Model of the Mg(OH)$_2$ Lattice (22).

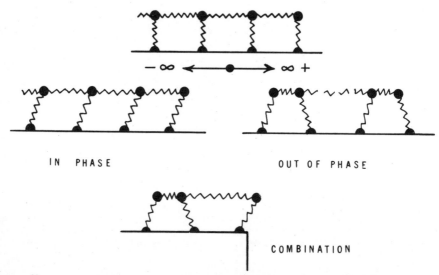

$- \infty \longleftarrow \bullet \longrightarrow \infty +$

IN PHASE

OUT OF PHASE

COMBINATION

MASS AND SPRING MODEL OF THE Mg(OH)$_2$ CRYSTAL SURFACE

Fig.3 Mass and spring model of the Mg(OH)$_2$ crystal surface
(22).

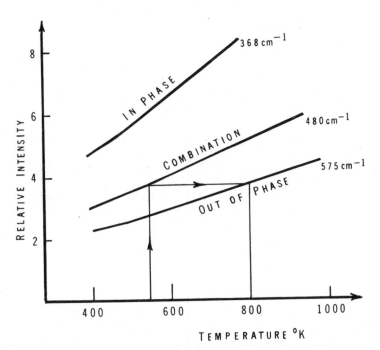

AMPLITUDE VERSUS TEMPERATURE FOR THE IN-PHASE, OUT-OF-PHASE, AND COMBINATION VIBRATIONS.

Fig.4 Amplitude versus temperature for the in-phase, out-of-phase, and combination vibrations.

113

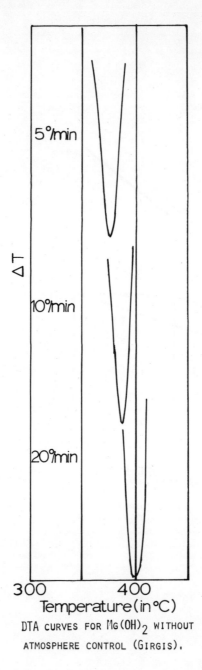

DTA curves for $Mg(OH)_2$ without
atmosphere control (Girgis).

Fig.5 DTA curves for $Mg(OH)_2$ without atmosphere control (23).

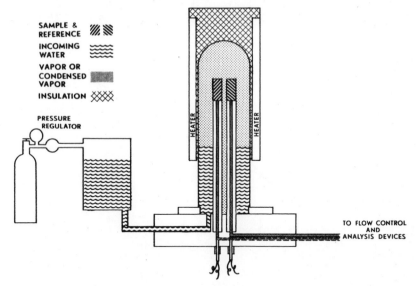

Fig.6 Furnace assembly for dynamic water vapor atmosphere (29).

Mg(OH)₂

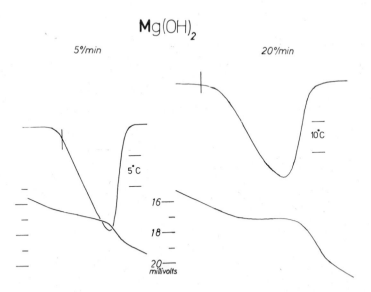

Fig.7 DTA and temperature curves for Mg(OH)$_2$ at a atm. (30).

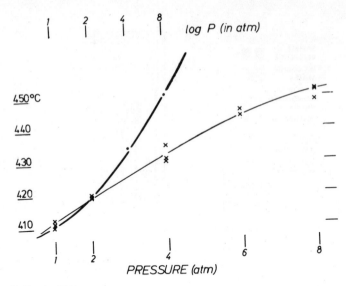

Fig.8 Peak DTA temperatures plotted vs. log P, ● , and vs. P,x (30).

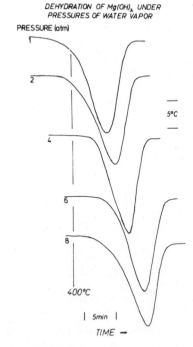

DEHYDRATION OF Mg(OH)₂ UNDER
PRESSURES OF WATER VAPOR

Fig.9 DTA peaks for $Mg(OH)_2$ at the indicated pressures and a heating rate of 5°/min. (30).

116

COMPLEMENTARY ROLE OF PRECISE CALORIMETRY AND THERMAL ANALYSIS

Hiroshi SUGA and Syûzô SEKI

Department of Chemistry, Faculty of Science
Osaka University, Toyonaka, Osaka 560

ABSTRACT

Several types of DTA apparatus were developed and used to survey the general feature of thermal behavior prior to the heat capacity measurement. Some examples which clearly show the usefulness of thermoanalytical technique in the early stage of thermodynamic investigation are presented. This technique is especially powerful to establish the phase relationship and its relative stability when the material produces metastable phases during thermal cycling. In the case of cyclohexene, the condition for preparing two kinds of metastable crystalline phases were clarified by use of DTA technique. Application of calorimetry to these metastable phases revealed that both phases have finite residual entropies and show respective glass transition phenomena. This provides the first example of existence of plural glass transition phenomena in one and the same material.

In the case of ethanol, frozen states of liquid phase and newly found crystalline phase were realized by a delicate control of cooling rate. Calorimetric measurements revealed the fact that the heat capacity data previously reported for glassy and supercooled states of liquid corresponded to those for the metastable crystalline phase. The heat capacity data for the glassy and the supercooled liquid of ethanol were determined for the first time. These experimental finding obtained by calorimetric investigations combined with DTA techniques suggest strongly that the familiar liquid-glass transition is just example of a class of transition which may be quite broad occurrence in condensed matter. The complementary role of qualitative and quantitative technique is emphasized throughout this report.

INTRODUCTION

The adiabatic calorimeter originally developed by Nernst is believed to yield the most accurate heat capacity data at low temperatures. The genuine calorimetry does require, however, a lot of sample with highest quality, laborious work with sophisticated techniques and enormously long time with continued patience. On the other hand, the thermoanalytical method has advantages in that it is simple, rapid ($10 \sim 10^2$ times more rapid than calorimetry) and is amenable to automatic recording over a wide temperature range with a minute amount of specimen ($10^{-2} \sim 10^{-3}$ times less than calorimetry). Therefore it is convenient and even necessary to survey the general features of thermal behavior of a particular meterial by thermoanalytical methods prior to the heat capacity measurement. The curve reveals immediately an approximate purity of the sample as well as the temperature and nature of phase transitions possibly occurring in the material. The preliminary investigation is especially fruitful to know the phase relationship and its relative stability when the material produces metastable phase during thermal cycling. Some examples which clearly shows the usefulness of thermoanalytical technique in the early stage of thermodynamic investigation are presented in this report, emphasizing the complementary role of both techniques.

APPARATUS

Three types of DTA apparatus were developed in our laboratory. The first is a standard type apparatus[1] which is possible to operate in the temperature range from 20 to 500 K. Sample is put into a glass tube provided with an inner, thin-walled capillary which housed a Chromel-P-Constantan thermocouple. The tube is evacuated and filled with helium as an exchange gas. A duplicate tube containing a suitable amount of alumina serves as a reference. Both tubes are placed in a copper block having two symmetrical cylindrical cavities to accomodate them, as shown in Fig. 1. The copper block is hung in a glass mantle connected with a vacuum apparatus. The temperature difference between the tubes is recorded in a standard fashion.

The second apparatus[2] is a modification of the first one to detect simultaneously dielectric loss of a specimen under an electric field of high frequency. The electrical energy absorbed by dielectrics changes into thermal energy which raises the temperature of the specimen and is easily detected by the usual DTA technique. Electrodes for applying the field are a pair of aluminium foils attached to the outer surface of glass tube containg a specimen. The electrodes are electrically insulated from the copper block by a thin sheet of polyethylene, which also serves to keep the electrode in place. Fine copper leads are pressed to the electrode beneath the polyethylene sheet and are connected to a high frequency oscillator.

The third one[3] is specially designed apparatus for realizing glassy state of a material as shown in Fig. 2. A copper block is joined to bottom of the Dewar vessel for liquid hydrogen. Sample and reference holders are made of a sheet of copper and the center of the sheet is soldered to the bottom of the block. A pair of constantan wires is soldered to the center of each holder so that they constitute a differential thermocouple. Sample vapor is deposited onto the sample holder via a needle valve which serves to control the deposition rate.

RESULTS AND DISCUSSION

One example of the dielectric loss measurement by DTA technique is shown in Fig. 3. A slight endothermic base-line shift of thermogram observed for glycerol corresponds to a glass transition. The dielectric loss peak as is manifested by the exothermic peak shift to the higher temperature with increasing frequency of applied field. A similar behavior is observed for cyclohexanol crystal. An interesting feature of this crystal is seen in the curve for 10.1 Mc obtained on cooling direction, in which the peak due to the crystallization is superposed on the higher temperature side of the dielectric loss peak without breaking the continuity of the latter. This provides evidence that the mode of the rotational motion of the molecules does not change essentially upon crystallization.

This new technique was also applied to the quinol methanol clathrate compound[4] in order to get an information of motional state of the guest molecule in the host lattice. This compound is considered as one of the representative which realizes the model of "particle in a potential box". The DTA recordings obtained on heating direction are shown in Fig. 4. The uppermost curve corresponds to the field-free experiment and shows an endothermic anomaly at around 66 K. The others represent the effect of the electric field at differnt frequencies. One remarkable point is that the abrupt increase of the dielectric loss takes place at the transition point. Motivated with these findings, the heat capacities of the compound were measured from 13 to 300 K. As is evident in Fig. 5, a sharp phase transition is observed at 65.74 K. The entropy of transition is (4.84 ± 0.2)J K^{-1} (mol of the trapped methanol)$^{-1}$ and this value corresponds approximately to R ln 2. It is conjectured that the dipolar interaction between the guest molecules might be responsible for alignment of their molecular axis head-to-tail, leading to the phase transition.

Cyclohexanol exhibits an interesting behavior in that

its high-temperature phase (crystal-I) can be supercooled easily and shows a heat capacity anomaly closely similar to that associated with a glass transition of a supercooled liquid. Kelley[5] measured the heat capacity and concluded that the supercooled crystal-I has the same value with that for low-temperature phase (crystal-II). On the other hand, Otsubo and Sugawara [6] found a relaxational nature of the heat capacity anomaly. This stands in contrast to the fact that the supercooled crystal-I is in the internal equilibirium state from the entropy consideration. They also reported the appearance of metastable intermediate phase during the course of irreversible transformation from the supercooled crystal-I to the stable crystal-II.

In order to clarify the origin of the discrepancy between these two results and also to study the thermodynamic properties of metastable crystal-III, low temperature calorimetry (7) was applied to this peculiar material. Fig. 6 shows the DTA curve under various conditions. By this survey, the condition of preparing the pure crystal-III was established. Heat capacity data of respective phases are plotted in Fig. 7 as functions of temperature. Crystal-II and crystal-III were found to obey the third law of thermodynamics from the comparison of calorimetric and spectroscopic entropies. While, crystal-I was found to remain a finite amount of entropy (4.72 ± 1.6)J K^{-1} mol^{-1} at 0 K. Also enthalpy relaxation phenomena of crystal-I were found around 150 K. The relaxation time τ was determined by analysing the spontaneous heating rate due to stabilization effect under adiabatic condition. The results are; τ/hr= 11.9 at 144 K, 32.3 at 141 K, 60.4 at 137 K and 125 at 133 K.

Dielectric investigation[8] for crystal-I revealed the existence of β-relaxation phenomenon at the temperature range far below the primary "T_g". It is interesting to note that the heat capacity of the frozen crystal-I starts to depart from that of crystal-II at around 75 K at which the relaxation time of β-relaxation becomes of the order of a few minutes, the time duration of each heat capacity measurement.

The heat capacity jump, the stabilization effect as well as the existence of residual entropy of crystal-I are characteristic properties of glass realized by cooling of liquid. We proposed a term "glass crystal" for the frozen crystalline state. In this connection we call the ordinary glassy state obtained from the supercooled liquid as "glassy liquid" in this text. Many examples of glassy crystal were found for alicyclic compounds[9] as shown in Fig. 8, as well as aliphatic compound like 2,3-dimethylbutane.[10]

In the case of cycloheptatriene and cycloheptane, direct transformation from the supercooled crystalline phase to the stable low-temperature phase, while the other compounds have intermediate metastable phases during the course of irreversible transformation. All these compounds have small entropy of fusion, usually less than 20 JK^{-1} mol^{-1}, and belong to the category of plastic crystal originally defined by Timmermans[11]. These experimental findings suggest that the familiar liquid-glass transition is just one example of a class of "transition" which may be of quite broad occurrence in condensed matter.

In the case of cyclohexene[12], glassy crystal-I state was realized by rapid cooling of the plastic crystalline phase. The supercooled crystal-I transforms on heating into the stable crystal-II by passing through an intermediate metastable crystal-III. Crystal-III was also found to be realized directly during the cooling of liquid with a moderate cooling rate. Heat capacity data for crystal-I, -II and -III are plotted in Fig. 9 in an enlarged scale. The big heat capacity jump around 81 K observed for crystal-I corresponds to T_g phenomenon. A small but definite heat capacity anomaly was observed for crystal-III around 83 K. The anomaly had a relaxational nature and was attributed to another T_g phenomenon intrinsic to crystal-III. We were able to measure the enthalpy relaxation associated with this glass transition and get a value for activation enthalpy kJ mol^{-1} which was different from that for crystal-I (18 kJ mol^{-1}). The residual entropy was determined calorimetrically as 11.7 JK^{-1} mol^{-1} for crystal-I and 2.6 JK^{-1} mol^{-1} for crystal-III, respectively.

As the heat capacity jump at T_g as well as the relaxation process of crystal-III were too minute to be detected by DTA, it was only possible to find out by use of the precise calorimetric method. On the other hand, the glass transition phenomenon(78 K) for glassy liquid of cyclohexene was only detectable by use of the vapor condensation type DTA apparatus which enabled an extremely rapid quenching. Cyclohexene offers the first example of plural glass transition phenomena for different phases of one and the same compound. Anyhow, it is a remarkable fact that these T_g values are not so different from each other in spite of the difference in their states of aggregation.

The enhanced role of DTA will culminate in the example of ethanol. Since the heat capacity measurement by Parks et al.[13] and Kelley[14], ethanol has been accepted for long time as one of the typical low molecular weight materials which form glassy liquids.

It was revealed by our recent investigation[15] that this widely-believed interpretation is not true in its original meaning and that the glassy liquid as well as glassy crystal states could be realized from the liquid by a delicate control of the cooling rate.

The DTA traces are reproduced in Fig. 10. Run 1 shows the heating curve of rapidly quenched sample from liquid with a cooling rate more than -50 K min^{-1}. T_g for glassy liquid is observed around 95 K. Run 2 shows the cooling curve from liquid with the cooling rate of -2 K min^{-1}. The supercooled liquid is found to crystallize into a new crystalline phase (crystal-II) around 125 K. Run 3 is for heating of crystal-II and clearly indicates the existence of glass transition phenomenon. Crystal-II melts at 127 K, but during the melting process the irreversible transformation into stable crystalline phase (crystal-I) occurs inevitably.

The precise heat capacity measurement was made from 13 to 300 K. In this case it was possible to prepare the sample of glassy liquid in the calorimeter cell *in situ* by rapid quenching of liquid with direct pouring of liquid nitrogen into the calorimeter can. Below the melting point of crystal-I, we were able fortunately to get heat capacity data of supercooled liquid down to 150 K. All the heat capacity data are reproduced in Fig. 11. It is clearly shown that the data just above T_g of glassy liquid are smoothly connected with the extrapolated line for the supercooled liquid. The data reported by the previous workers are in agreement with our present data for glassy crystal, and not the glassy liquid. It is presumably an undreamt-of concept at that time to imagine that a crystal may have a kind of glass transition phenomenon.

During the course of study on vitrification process of *iso*-propylbenzene from liquid, an interesting phenomenon was observed[16] by DTA method. Fig. 12 reproduces the observed thermogram. Run 1 is for a cooling with the rate of -0.7 K min^{-1}. A sharp exothermic peak is observed at 102 K after the vitrification sets in. This sharp peak is associate with the formation of cracks in the sample at glassy state. The heat evolution is accompanied with sounding and flashing of visible light as described below. Run 2 is for the heating and only a peak due to T_g is observed. Run 3 is for the heating of glassy sample which is subjected to annealing at 150 K. The endothermic peak at 176 K corresponds to the fusion of the crystal induced during the annealing.

The luminescence, accompanied with heat evolution

and sound crack, is so weak in its intensity that it is only visible in a complete darkness. These phenomena remind us of triboluminescence. Although a detailed mechanism is not clear, the following explanation may be a possible one. Quenching the sample causes it to be strained because it is difficult for molecules to readjust themselves after the sample vitrifies. Cracks liberate distortion energy and bring about triboelectricity between the glass wall and the sample or inside the sample. Luminescence may be thought as being due to the discharge of triboelectricity. In either event, the accumlated distortion energy is released as the combined forms of thermal, sound and light energies.

The heat capacity curves[17] of *iso*-propylbenzene are given in Fig. 13. The primary glass transition occurs around 126 K with a drastic heat capacity change. Exothermic temperature drift during the equilibration periods of heat capacity measurement are observed around T_g. The observations of enthalpy relaxation are practically limited within a temperature interval, usually $10{\sim}20$ K below T_g, owing to the prolonged relaxation time over the usual laboratory time scale. Unusual temperature drifts and a step-like heat capacity increment were encountered during the course of heat capacity measurement at around 70 K, far below T_g. Slight exothermic temperature drifts appeared at about 50 K and gradually increased up to 70 K above which the drifts changed into endothermic ones. This behavior of temperature drift is quite similar to those found in the primary T_g, albeit it is very small effect.

Johari and Goldstein[18] found the β-relaxation for *iso*-propylbenzene by means of dielectric measurement and showed that the Arrhenius plot of the frequency of maximum loss $f$max against $1/T$ gives a straight line. Extrapolating their data to lower temperature, one can get a temperature 76 K at which $f$max becomes equal to 10^{-3} Hz. This time scale corresponds approximately to the time duration of the heat capacity determination. The calorimetric β-transition temperature of about 80 K, is in good agreement with that inferred from the dielectric measurement. This agreement is reasonable in view of time-temperature equivalence of the dispersion effect and suggests that these anomalies are the calorimetric manifestation of the β-relaxation phenomena.

CONCLUSION

Low-temperature calorimetry was developed by Nernst about 70 years ago, while, thermoanalytical techniques have a longer history. Nevertheless, these useful techniques have scarcely been utilized by calorimetrist without obvious reason. The information output derived from calorimetry can undoubtedly be enchanced by the thermoanalytical survey carried out in the early stage of investigation. The qualitative and the quantitative investigations constitute two important aspects of key to clarify the nature. A wagon can move smoothly only when the wheels of both sides work well.

REFERENCES

(1) H.Suga, H.Chihara and S.Seki, J.Chem.Soc. Japan, Pure Chem. Sect. (Nippon Kagaku Zasshi), 82, 24 (1961)
(2) T.Matsuo, H.Suga and S.Seki, Bull. Chem. Soc. Japan, 39, 1827 (1966)
(3) O.Haida, H.Suga and S. Seki, Thermochimica Acta, 3, 177 (1972)
(4) T.Matsuo, H.Suga and S.Seki, J. Phys. Soc. Japan, 22, 677 (1967); Idem, Proc. 1st Intern. Conf. Calorimetry and Thermodynamics, Warsaw 1969
(5) K.K.Kelley, J. Am. Chem. Soc., 51, 1400 (1929)
(6) A.Otsubo and T.Sugawara, Sci. Rep. Res. Inst. Tohoku Univ., A7, 583 (1955)
(7) K.Adachi, H.Suga and S.Seki, Bull. Chem. Soc. Japan, 41, 1073 (1968)
(8) K.Adachi, H.Suga, S. Seki, S.Kubota, S.Yamaguchi, O.Yano and Y.Wada, Mol. Cryst. Liquid Cryst, 18, 345 (1972)
(9) K.Adachi, H.Suga and S.Seki, Bull. Chem. Soc. Japan, 43, 1916 (1970)
(10) K.Adachi, H.Suga and S.Seki, Bull. Chem. Soc. Japan, 44, 77 (1971)
(11) J.Timmermans, J. Phys. Chem. Solids, 18, 1 (1961)
(12) O.Haida, H.Suga and S.Seki, Chem. Letters, 1973, 79
(13) G.E.Gibson, G.S.Parks and W.M.Latimer, J. Am. Chem. Soc., 42, 1542 (1920); G.S.Parks, J. Am. Chem. Soc., 47, 338 (1925)
(14) K.K.Kelley, J. Am. Chem. Soc., 51, 779 (1929)
(15) O.Haida, H.Suga and S.Seki, Proc. Japan Acad., 48, 683 (1972)
(16) K.Kishimoto, H.Suga and S. Seki, J. Non-cryst. Solids, in press (1974)
(17) K.Kishimoto, H.Suga and S.Seki, Bull. Chem. Soc. Japan, 46, 3020 (1972)
(18) G.P.Johari and M.Goldstein, J. Chem. Phys., 53, 2372 (1970)

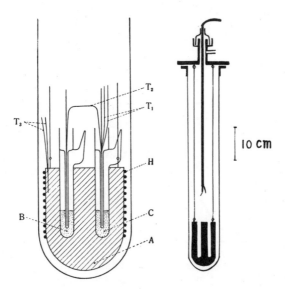

Fig. 1   DTA apparatus developed for low temperature study.

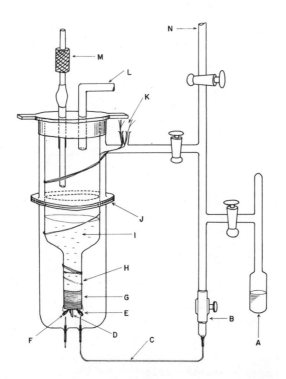

Fig. 2   Vapor deposition type DTA apparatus.

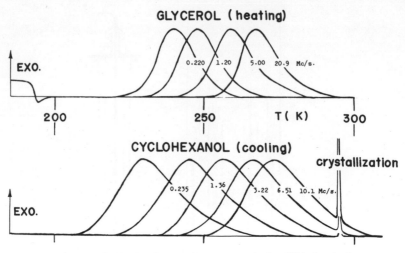

GLYCEROL (heating)

EXO.

0.220 1.20   5.00 20.9 Mc/s.

200   250   T (K)   300

CYCLOHEXANOL (cooling)

crystallization

EXO.

0.235   1.36   3.22 6.51 10.1 Mc/s.

Fig. 3   Dielectric loss measurement by DTA technique.

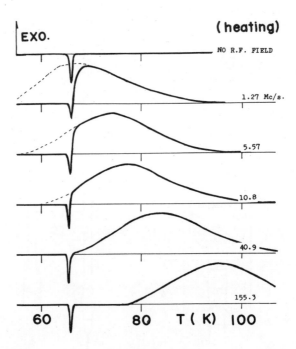

EXO.   (heating)

NO R.F. FIELD

1.27 Mc/s.

5.57

10.8

40.9

155.3

60   80   T (K) 100

Fig. 4   Thermogram of quinol methanol clathrate compound
         under electric field.

126

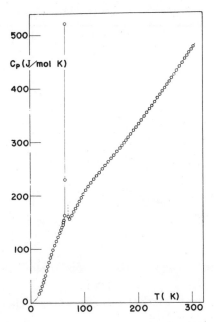

Fig. 5   Heat capacity curve of quinol methanol clathrate
compound.

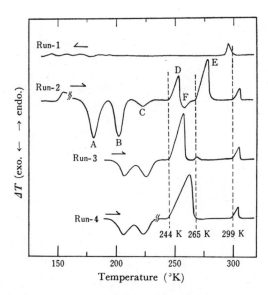

Fig. 6   DTA curves of cyclohexanol under various conditions.

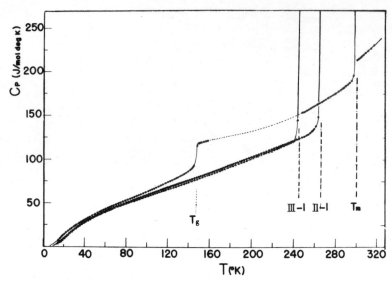

Fig. 7   Heat capacity curve of cyclohexanol.

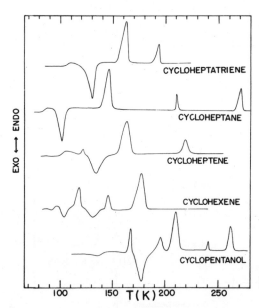

Fig. 8   DTA curves of quenched plastic crystal.

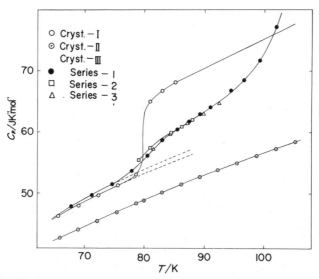

Fig. 9　Heat capacity curve of cyclohexene.

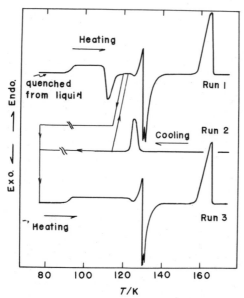

Fig.10　DTA curves of ethanol under various conditions.

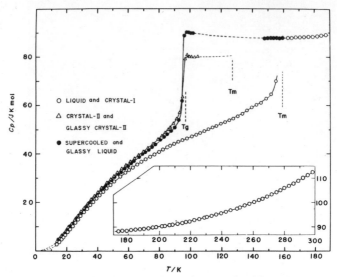

Fig.11   Heat capacity curve of ethanol.

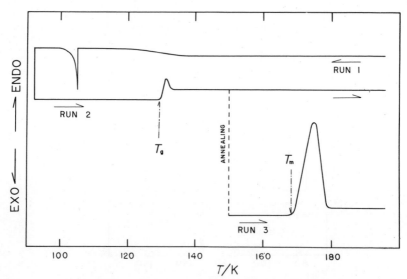

Fig.12   DTA curves of *iso*-propylbenzene.

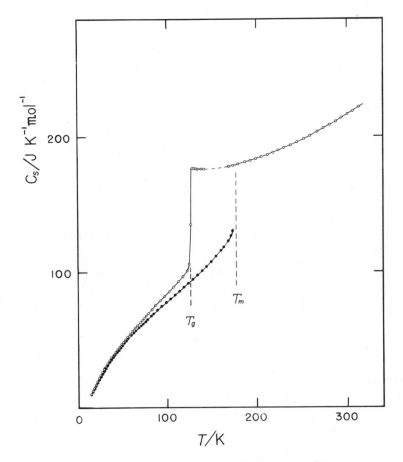

Fig.13   Heat capacity curve of *iso*-propylbenzene.

THERMOGRAVIMETRY OF ELASTOMERS

W. R. Griffin

Air Force Materials Laboratory
Wright-Patterson Air Force Base, Ohio, USA

ABSTRACT

Thermogravimetry (TG) can be used to rank materials in order of increasing service capability in high temperature environments TG is particularly well suited for evaluating elastomers because the predominating mechanisms of thermal degradation involve weight losses The temperature for onset of weight loss has been a very useful figure, indicating the upper limit of high temperature capability. The value of the data obtained depends upon reliable instrument behavior, careful preparation of materials, knowledge of sample history, purity of purge gases, etc. Interpretation of a TG curve for onset temperature requires insight into elastomer composition, mechanisms of degradation, and careful scrutiny of the complete weight loss curve. A valuable ranking of new materials as well as an upper thermal limit can be obtained by TG even though the realistic long term service temperature will be lower due to the added features of the actual environment.

INTRODUCTION

Elastomeric components (rubber parts) are very often the service life limiting items in today's machines, vehicles, and appliances. This need for new polymers and improved formulations generates a need to sort the serviceable from the unserviceable. The actual sorting and ranking procedure varies with the inherent properties of the polymer and the intended use of the final product. There are many methods of ranking, but the most efficient, in terms of time and sample requirements, are based on the results from analytical instruments designed to evaluate a specific feature and render accurate, simple data.

The Air Force Materials Laboratory has particular interest in high temperature resistant elastomers. High performance aircraft have operational temperatures that are above the long life[*] capability of tires, hose, seals, fuel tank sealants, wire insulation, rain erosion, coatings, etc   The ability of an elastomeric material to withstand high temperatures can be determined

---

[*] Long term service is in the order of 1,000 to 10,000 hours of vehicle operation.

132

by measuring changes in a variety of properties, including tensile strength, modulus, elongation, hardness, weight, resilience, shape, volume, tear strength, chemical composition, color and surface features. However, many of these changes are interdependent, of no consequence, or do not yield consistent information. The most consistent, meaningful behavior of elastomeric materials at degradation temperatures is weight loss. Therefore, a means of detecting onset of weight loss versus temperature would provide a single temperature value for each material, i.e., 140°C for poly(ethylene-co-propylene) elastomer in air. While several complicating effects interfere with the credibility of the onset temperature as a measure of upper use limit, it is a reasonable method for ranking new materials for service in high temperature environments.

The following description of a reliable thermogravimetry (TG) instrument, some insights into polymer degradation mechanisms, and interpretations of TG curves are discussed to support the validity of onset of weight loss temperatures as the high temperature limit of elastomer materials.

EXPERIMENTAL

An early Stone DTA-TG instrument was modified to indicate zero weight, weigh the samples, and read out the data as percent weight loss versus temperature. Figures 1-4 are photographs of the parts discussed in the text below. The weighing device is a calibrated spring suspended from the non-rotating member of a micrometer lift screw. The tension on the spring results from the cumulative weight of the spring, slotted flag, connecting tube, crucible, and sample. The location of the lower end of the spring is sensed by two fixed photocells which are illuminated by a fixed light source through the slot in the suspended flag. Unequal illumination of the photocells results in a signal which causes rotation of the servomotor which is geared to the shaft of the lift screw. This raises or lowers the spring and attached components as required to maintain the balanced condition. The overall result is a rotation of the servomotor that is directly proportional to the change in weight while the sample crucible remains essentially stationary.

A ten-turn adjustable resistor, geared to the servomotor, is part of a voltage dividing network with other adjustable resistors and a battery, as shown schematically below.

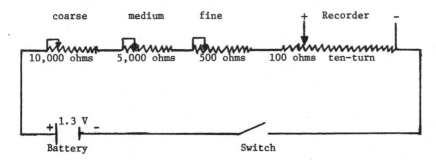

The output to the recorder is zero with the empty crucible because the ten-turn resistor is at the low limit (to the right in the schematic). A sample weighing more than 0.475 grams, with the other adjustable resistors maximum to the left, will cause the output to exceed the 10 millivolt limit of the recorder. Therefore, a 0.4 gram sample is roughly weighed to the nearest ±0.05 gram. This is a good sample size and does not lose zero index by overdriving the ten-turn resistor. The other adjustable resistors (coarse, medium, and fine) can be lowered in value to cause 10 millivolts to be delivered to the recorder. Since zero output has not changed by this scale expansion, the readout is zero to one hundred percent weight loss. The form of the readout is illustrated by Figure 5.

The preferred sample form is vulcanized strip 5mm x 20mm x 1-2mm thick. Polymer crumbs can be used if melting does not occur before thermal degradation (the combination of purging gas and the wire mesh bottom of the crucible causes weight loss by dripping if melting occurs). The sample size can be 0.1 to 0.45 grams. The larger size is preferred because the increased span of the ten-turn readout resistor minimizes signal error.

Care is taken to prepare samples free from mold release and other contaminants. Samples already in the vulcanized form are selected to be free from contamination, i.e., center cuts from the elastomer item. Purge gases are prepurified helium or nitrogen. Air directly from the laboratory is pumped for the air runs (relative humidity varies with the seasons). In general, two TG runs are made on each elastomer composition; one in an inert dry gas, and the other in ambient air. Two milliliters per minute of gas flow through the crucible and furnace. The 1°C/min. heating rate provides a near-equilibrium thermal environment.

RESULTS AND DISCUSSION
Mechanisms of elastomer degradation

General use of the term "Elastomer" refers to a composite of elastomeric polymer, crosslinked into a network, with reinforcing fillers, crosslinking agent residues, inhibitors, plasticizers, antioxidants, etc. Serviceability depends upon the retention of mechanical properties which, in turn, depends upon the three dimensional elastomeric network. Since the network is polymeric, a knowledge of polymer degradation mechanisms, as well as interaction with other agents of the composition, is important (1,2).

Three (3) general mechanisms describe most of the observed modes of polymer degradation at high temperatures. The most frequently encountered is the loss of small molecules followed by charring. The second in frequency of occurrence appears to be a breakdown into volatile fragments, leaving no residue, often preceded by loss of small molecules. The third is unzipping that regenerates the monomer from which the polymer was made. Included in this latter mechanism are special cases where several monomer units or a specific molecule is released from the polymer chain.

A good indicator of degradation mechanism is the shape of the TG curve. The gradually increasing rate of weight loss indicates the loss of small molecules. The change from small molecule loss to char is often gradual, but the slope change indicates the slower process for small molecule release from char. The elimination of fragments most often begins by the release of small molecules. The transition to fragments depends upon the polymer system. The more labile structures are eliminated first, possibly at a high rate, then the more stable structure is broken down at a higher temperature and usually at a slower rate. Complete volatilization of the polymer part of an elastomer below 600°C is a good indication that carbon char was not formed. The onset of rapid weight loss with no change in rate until the polymer part is volatilized indicates unzipping. The monomer is usually evolved but may react before escaping the hot crucible. The examples below illustrate the shape of the TG curve for a specific mechanism.

Small molecule elimination with charring

Poly(chlorobutadiene) thermally degrades by the release of hydrogen chloride and char formation. Figure 6 illustrates the slowly increasing rate of weight loss to the point of 40% loss. The hydrogen and perhaps some methane, ethylene or higher homologs continue to evolve as the char increases in carbon content. Finally, the char volatilizes, leaving an empty crucible at 700°C. Trace quantities of water in the water pumped helium remain after passing through a drying column and may account for the loss of high carbon char below the vaporization temperature of carbon.

Fragmentation

Poly(vinylidene fluoride-co-hexafluoropropylene) undergoes release of hydrogen fluoride followed by fragmentation. Figure 7 shows TG of a gum vulcanizate containing hexamethylenediamine-derived crosslinks and 12.8% (15 parts with 100 parts polymer) of magnesium oxide. The hydrogen fluoride reaction with the oxide releases water and appears as a very slow weight loss in the 200-300°C region. Above 400°C the release of hydrogen fluoride reaches a high rate. Since the molar ratio of hydrogen to fluoride in the copolymer is approximately 9:15, there is insufficient hydrogen to cause all of the fluorine to be released as hydrogen fluoride. In addition, volatile fragments contain some hydrogen. The conclusion is that hydrogen fluoride and fragments are eliminated to 450°C where a high carbon network begins to break down into the final volatile fragments. The conversion of 12.8% MgO to MgF_2 accounts for the 22% inorganic residue (theory requires 26% residue).

Unzipping

Poly(tetrafluoroethylene-co-trifluoronitrosomethane) degrades by unzipping after some initial loss of small molecules. Figure 8 shows the onset of continuing weight loss at 200-220°C with unzipping proceeding rapidly at 260°C. The compound contains 11% inorganic ash, which is all that remains above 320°C. Poly(perfluoroalkylene-triazine) is another good example of degradation by the unzipping mechanism. Triazine rings cleave

into the fluorocarbon nitriles from which they were formed.
Figure 9 shows 2% loss to 430°C where rapid weight occurs. The
initial 2% is most likely low molecular weight polymer vaporiz-
ing at this high temperature. The five parts of iron powder
used as a crosslinking catalyst is the residue at 550°C.

The temperature range of the TG curve where the mechanism
is apparent is beyond the useful limit of the elastomer. The
network is rapidly being destroyed. The value of the whole TG
curve is to reveal the mechanism and other oddities that can be
translated into an interpretation of the initial loss, or onset
of weight loss. This is in the temperature range where the network is just beginning to break down. This is the upper temperature limit of the elastomer for processing or service. Therefore,
the onset of weight loss (of the network) is a critical temperature for the elastomer and can be used with judgment as the upper
limit of high temperature capability. A logical procedure to
judge this onset temperature involves TG runs with air and inert
gas purge on a series of compositions of the elastomeric polymer
In progressing from pure polymer, to gum vulcanizate, to reinforced vulcanizate, to that modified by plasticizers, antioxidants, inhibitors, etc., the true onset of network degradation
can be judged.

Interpretation of selected thermograms

The TG curves discussed below have been selected from the
many that have been run on new materials developed by or sub-
mitted to the Air Force Materials Laboratory for consideration
in high temperature environments. In addition to reproductions
of the original thermograms, an interpretation is offered that
is based upon insight into the material and instrument behavior.

Fluoroalkoxy phosphazene elastomers

Figures 10 through 14 are thermograms of poly(fluoroalkoxy
phosphazene) elastomers with the polymeric structure shown below.

$$\left[\begin{array}{ccc} CF_3 & & (CF_2)_3CF_2H \\ CH_2 & & CH_2 \\ O & & O \\ P \!\!=\!\! N \!\!-\!\! & P \!\!=\!\! N \\ O & & O \\ CH_2 & & CH_2 \\ CF_3 & & (CF_2)_3CF_2H \end{array}\right]_n$$

Figure 10 shows a 1% loss of volatiles during heating to
220°C. This is interpreted to be low molecular weight polymer,
moisture, and possibly some hydrolysis products from months of
ambient storage. Increasing weight loss sets in near 230°C and
maximizes at 340°C. The maximum rate shown by the derivative
curve to the right of the weight loss curve has little significance to long term service conditions. It shows decomposition
of the elastomer in minutes. The departure of the weight loss

curve from the zero rate near 230°C indicates the upper use temperature, above which the rate of degradation precludes most useful application of the elastomer

The companion run in air environment shown in Figure 11 has very similar behavior. This polymer is very resistant to oxidation and, in addition, the mode of degradation does not leave an oxidizable site in the chain. Release of large molecules (most likely as the alkoxy phosphites or phosphates) explains the absence of inorganic phosphorous residue. Both thermograms show the approximate 30% ash, which is the amount of inorganic reinforcement used in the elastomer.

Figure 12 illustrates loss of moisture as well as volatiles resulting from degradation during exposure to 95% relative humidity (RH) at 93°C. Interpretation of the 4.5% loss at 150°C as moisture absorbed in the humidity chamber is modified by the 31% inorganic ash remaining. If the curve were adjusted to 100% weight at 150°C, the ash content would be higher than the actual 31% of the original helium run. Since the ash is correct on this post humidity run, we conclude that the loss is caused by the same products (moisture, low molecular weight polymer and degradation products) that were released during the helium run. The loss has increased 3.5% indicating hydrolytic degradation of the elastomer.

Figure 13 is the helium purged base-line thermogram of an improved version of poly(fluoroalkoxy phosphazene) with stability to 200°C and 36% ash. Figure 14 shows the effects of 95% RH at 93°C for 28 days. The 2% loss at 200°C appears to be at least 1.5% hydrolysis products because the helium TG curve showed about 0.5% loss at 200°C and the ash contents are very close.

The conclusion is that poly(fluoroalkoxy) phosphazenes are thermally stable to 200°C and have sufficient resistance to high humidity environments to be considered serviceable. The degradation mechanism appears to be the special case of unzipping where specific molecules (not the monomer) are released.

Poly(ethylene-co-propylene) elastomer

Figure 15 shows the effect of oxidation on the vulcanizate below:

Poly(ethylene-co-propylene)	100.0
Poly(chlorobutadiene)	8.0
Irganox* 1076	0.6
Dilaurylthiodipropionate	0.3
Zinc Oxide	14.0
Magnesium Oxide	5.0
SAF Black	50.0
Dicumyl Peroxide	3.5

Press cured 60 minutes @ 160°C.

* Registered Trade Mark of Geigy Chemical Corporation, Ardsley, New York.

The weight loss up to 100°C can be attributed to moisture and outgasing of peroxide decomposition products. The region from 140 to 210°C is difficult to assess because the antioxidant (Irganox 1076) and stabilizer (dilaurylthiodipropionate) can be volatilized and oxidized. A hint of volatilizing occurs at 140°C with slow oxidation to about 210°C. Above 210°C, the rate of weight loss is progressively increasing with temporary breaks around 420 and 450°C as the char and carbon black filler oxidize at their particular rates. Finally, at 620°C only the inorganics remain. The 6% compares well with the 10.5% inorganic of the compound if one considers that ZnO becomes the more volatile $ZnCl_2$.

We conclude from the above that long term service life in air would be limited to 140°C and that above 210°C the life would be very limited. Aging this compound in forced circulation ovens confirms this conclusion of significant deterioration in the 200°C region (3).

	Original	8 Hours @ 204°C	24 Hours @ 204°C
Tensile, psi	3180	2655	2410
Elongation, %	375	375	305
Hardness, Shore A	69	86	86

The progressive reduction of tensile strength and elongation with increasing hardness indicates continual loss of performance capability with time at 204°C.

Poly(trifluoropropyl, methyl, siloxanes)

Figure 16 shows the excellent thermal stability of the fluorosilicones. No weight loss is noted up to 240°C, then a smooth weight loss curve to 34% ash at 600°C. The formulation below was used as a standard for comparing many fluorosilicone variations. It has the bare minimum of filler and stabilizer, which emphasizes the difference between the polymer variations.

Poly(trifluoropropyl, methylsiloxane)	100
Fumed Silica	20
Iron Oxide	2
α2, α2'-bis(t-butylperoxy) diisopropylbenzene	1.5

Press cured 30 minutes @ 165°C;
Post cured 24 hours @ 204°C.

Postcuring the samples rids the compound of moisture and peroxide decomposition products as shown by the zero weight loss to 240°C. The weight loss curve shows a gradually increasing rate of loss above 240°C indicating that decomposition is not of unzipping mechanism. The 34% ash does not compare well with 18% expected. We conclude that the post cure was sufficient to rid the vulcani zate of peroxide decomposition products, that the polymer does not degrade by a simple unzipping but some condensing to more stable structure occurs, and that a portion of this condensed structure remains as ash. Figure 17 shows a higher rate of weight loss in the 300°C region for the same elastomer in an air environment. The 0.5% loss to 250°C is due to instrument warm-up and is not significant. The ash content of 34% confirms the helium TG run for the inorganic condensed structure.

We conclude that fluorosilicones are thermally and oxida-tively stable to the 240-250°C range, and that degradation is a combination of cyclic oligomer release and loss of alkyl groups from silicon which leads to inorganic condensed structure of about 16% (of the formulation).

Figure 18 is a thermogram of the fluorosilicone polymer when formulated as fuel tank sealant. The polymer structure is the same as in the foregoing figure except a viscous liquid polymer is chain extended and crosslinked by silicon-carbon bonding. In addition, the formulation has been optimized for resistance to a high temperature fuel environment. A gradual weight loss occurs from room temperature to 230°C and smoothly increases to about 380°C where a very rapid weight loss occurs. The ash content of 17% corresponds to the 17.5% inorganic filler. The initial gradual weight loss can be attributed to low molec-ular weight materials including water, gases, and low weight silicone polymers. More of its weight is retained at 300°C than for the base polymer formulation in Figure 16. In addition, the rate of loss at 300°C is much less for the optimized sealant. However, the rate of 400°C is higher for the sealant. The polymer backbone has silicone-carbon bonding as well as other ingredients to prevent early degradation. The result appears to be a change from partial release of cyclic oligomers to complete loss of polymer by this special unzipping mechanism.

Figure 19 shows side by side thermograms of a hybrid fluoro-silicone-fluorocarbon elastomer of the following structure and formulation:

$$\left[\begin{array}{cc} CH_3 & CH_3 \\ -SiCH_2CH_2CF_2CF_2CH_2CH_2Si-O- \\ CH_2 & CH_2 \\ CH_2 & CH_2 \\ CF_3 & CF_3 \end{array} \right]_n$$

Hybrid Fluorosilicone	100
Fumed Silica	20
Iron Oxide	2
$\alpha 2,\alpha 2'$-bis(t-butylperoxy) diisopropylbenzene	1.5

Press cured 30 minutes @ 165°C;
Post cured 24 hours @ 204°C.

The TG run with helium purge indicates thermal stability to 270°C, followed by rapid loss to 21% ash. In an air environment, oxidation sets in at about 130°C after the 0.5% instrument drift during warm-up. The degradation curve is similar in shape but begins 10°C sooner and has only 18% ash at 500°C.

We conclude that the polymer is being oxidized in the 130 to 260°C region, and that some char, perhaps silicon carbide, results during degradation in inert gas. The 18% ash of the run in air agrees well with 18% expected. Therefore, the char is oxidized by the air.

Figure 20 is a side by side thermogram of the same formulation to which has been added two parts of Irganox 1010 antioxidant prior to vulcanization. The thermal stability in helium is less if one considers that weight loss sets in about 10°C sooner, but greater if the 1% weight loss at 300°C is compared with the 2% of the formulation without antioxidant. The results in an air environment show the beneficial effect of the antioxidant. The onset of weight loss is at a slightly higher temperature and the rate of loss is lower. The difference is noted at 300°C where the unprotected formulation has lost 6% compared to 3% for the one with antioxidant.

We conclude that hybrid polymer is susceptible to oxidation in air above 130°C, that the mode of degradation is the release of chain fragments with little char formation, and that nonvolatile antioxidants can extend the useful life in an air environment.

Poly(tetrafluoroethylene-co-perfluoromethylvinyl ether) elastomer

Figure 21 shows the high thermal stability of the completely fluorinated polymer as a gum vulcanizate. Weight loss at 300°C is only 1% and most likely caused by release of low molecular weight polymer of this early sample. At 380°C, the weight loss is faster than the servomotor can balance. The stepping of the curve strongly suggests a melting (dripping) of the polymer from the crucible. The complete loss of total weight is characteristic of fully fluorinated polymers, but since the elastomer appears to degrade by melting, the amount of residue is uncertain.

We conclude that the elastomer appears to be stable to 280°C and that the mechanism of degradation is either cleavage of crosslinking sites followed by melting or chain scission to a melt. As this elastomer becomes optimized, this ambiguity can be resolved.

Poly(tetrafluoroethylene-co-trifluoronitroso methane) elastomer

This oxidation resistant, non-burning copolymer, already discussed as an example of unzipping, is crosslinked at pendent acid groups by chromium trifluoroacetate (5 parts with 100 parts copolymer) and reinforced with 10 parts fumed silica. Assuming that the chromium ash will be Cr_2O_3, the residue should be the

11% shown. The thermogram of Figure 8 shows excellent thermal stability at the 200°C range. The gradual loss under 200°C can be caused by loss of lower molecular weight polymers in this early production batch. The rapid weight loss above 260°C suggests unzipping of the monomers, but the products are not the monomers or fragments of the polymer containing several units, with or without small molecules eliminated. It appears that the polymer chain is unzipping and the highly reactive monomers are rearranging into the observed products -- carbonylfluoride, COF_2; perfluoro-N-methyleneimine, $CF_2=N-CF_3$ or isomer $CF_3CF=NF$; trifluoromethyl isocyanate, $CF_3-N=C=O$; carbon dioxide, CO_2; hydrogen fluoride, HF; silicon tetrafluoride, SF_4 and perfluorodimethylamine, $(CF_3)_2NH$ (4,5).

Poly(dimethyl siloxane-carborane) elastomer

The usual heat resistance claimed for this elastomer system (stability in the 370-480°C range) appears to be due to the solid degradation products. Figure 22 shows no significant weight loss in helium up to 300°C. A 2% loss occurs in the next 50-60°C, and about 10% is lost over the next 100°C. Above 600°C (12% loss), no further change in weight occurs. Figure 23 shows a slight reduction in the temperatures for the oxidation run but a similarly shaped curve.

We conclude from these thermograms that only two decompositions take place: one at 300°C, and one at 500°C. Knowledge of the polymer system allows us to conclude further that the dimethyl siloxane structure degrades at 300°C and the carborane at 500°C. The residue consists of inorganic borosilicates. This high ash content appears to be characteristic of this polymer system. Figure 24 is a TG curve of a fluorosilicone version of the siloxane carborane of the following structure:

$$
\left[
\begin{array}{c}
CH_3 \\
-SiCH_2CH_2 \quad —C \quad \overset{\displaystyle \setminus \bigcirc /}{\underset{B_{10}H_{10}}{}} \quad C—CH_2CH_2Si-O- \\
CH_2 \\
CH_2 \\
CF_3
\end{array}
\qquad
\begin{array}{c}
CH_3 \\
\\
CH_2 \\
CH_2 \\
CF_3
\end{array}
\right]_n
$$

It is shown only to confirm this high ash content. The thermogram shows 56% residue when the silicon and iron oxide inorganics amount to only 11%.

Onset temperatures for elastomers

The temperatures at which the onset of weight loss occurs for the elastomers discussed in this paper appear in Table 1. The onset temperature is the upper temperature that can be considered for processing or short term service. Below the onset temperature, the service life is very dependent upon the actual environment. The real environment contains agents that cause swelling or chemical deterioration as well as physical effects such as compression, flexing, or abraiding that greatly shorten the service life. As a rule, these other aspects of the environment become more severe as the temperature increases, requiring a careful simulation of the service environments. In spite of the ideal environments of TG, valuable data is obtained which allows rapid screening of emerging materials.

TABLE 1

Onset of weight loss temperatures
for various elastomers

| Material | Thermal Stability, °C. | |
	Helium	Air
Poly(chlorobutadiene) gum	195	---
Poly(chlorobutadiene) elastomer	170	---
Poly(vinylidenefluoride-co-hexafluoropropylene) elastomer	240	---
Poly(fluoroalkoxy phosphazene) elastomer	220	210
Poly(ethylene-co-propylene) elastomer	---	140
Poly(trifluoropropyl, methyl siloxane) elastomer	250	220
Poly(trifluoropropyl, methyl, siloxane) fuel tank sealant	---	220
Fluorosilicone-fluorocarbon hybrid elastomer	275	130
Poly(tetrafluoroethylene-co-perfluoromethylvinylether) elastomer	280	---
Poly(tetrafluoroethylene-co-trifluoronitrosomethane) elastomer	200	---
Dimethyl siloxane-carborane elastomer	300	290
Poly(perfluoroalkylene, perfluoroalkyl, triazine) elastomer	310	---

REFERENCES

1. K. L. Paciorek, R. H. Kratzer, and J. Kaufman, "Oxidative
   Thermal Degradation of Polytetrafluoroethylene," J. Polymer
   Sci. (Polymer Chemistry Edition), Vol II, 1465-1474 (1973)

2. R. T. Conley, "Thermal Stability of Polymers," Marcel Dekker,
   Inc., New York, Vol I (1970)

3. K. Murray, "Thermally Stable Ethylene Propylene Elastomers,"
   AFML-TR-71-25 (November 1971)

4. J. D. MacEwen and E. H. Vernot, "The acute Toxicity of Thermal
   Decomposition Products of Carboxy Nitroso Rubber (CNR),"
   Proceedings of the 4th Annual Conference on Atmospheric Contamination in Confined Spaces (Paper No. 19 of AMRL-TR-68-175)
   (10-12 September 1968)

5. G. A. Kleinberg and D. L. Geiger, "Tandem Thermogravimetric
   Analyzer - Time of Flight Mass Spectrometer System Designed
   for Toxicological Evaluation of Nonmetallic Materials,"
   Thermal Analysis (ICTA 1971) Vol I, 325-336 (AFML-TR-71-71)
   (October 1971)

Fig. 1 Disassembled TG instrument   Fig. 2 Photocell position detector

Fig. 3 Sample holder and furnace   Fig. 4 Assembled weighing mechanism

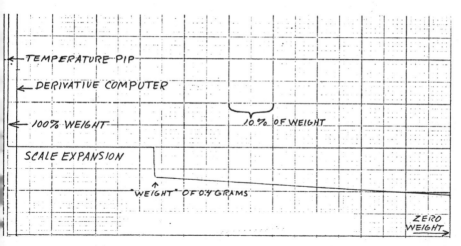

Fig. 5 Recording of data

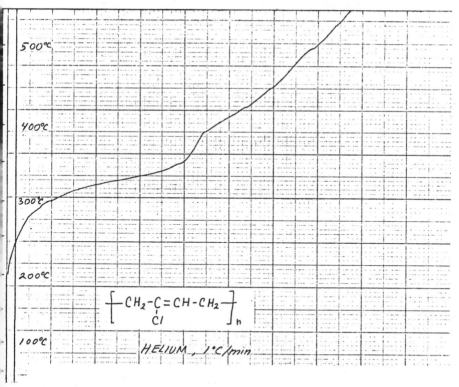

Fig 6 TG of poly(chlorobutadiene) in helium

145

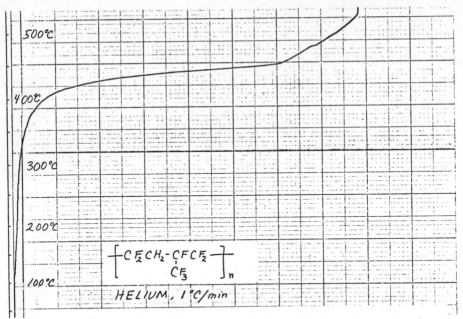

Fig. 7 TG of poly(vinylidene fluoride-co-hexafluoropropylene) elastomer in helium

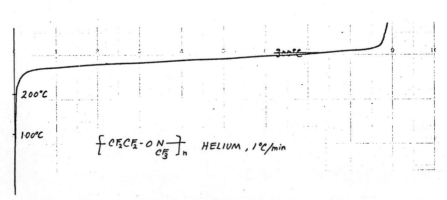

Fig. 8 TG of poly(tetrafluoroethylene-co-trifluoronitroso-methane elastomer in helium

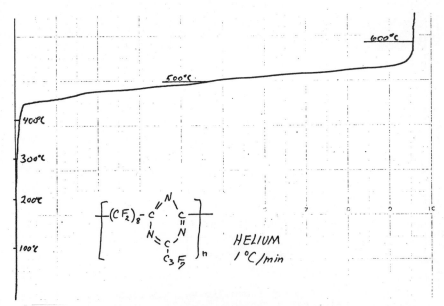

Fig. 9   TG of poly(perfluoroalkylene-perfluoroalkyl triazine
elastomer in helium

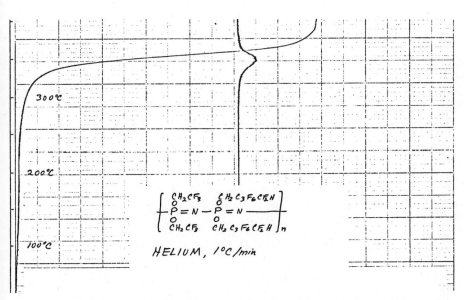

Fig. 10   TG of poly(fluoroalkoxy-phosphazene) elastomer in
helium, early version

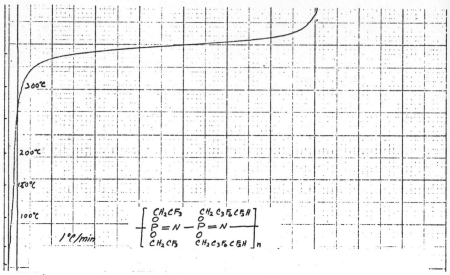

Fig. 11   TG of poly(fluoroalkoxy-phosphazene) elastomer in
air, early version

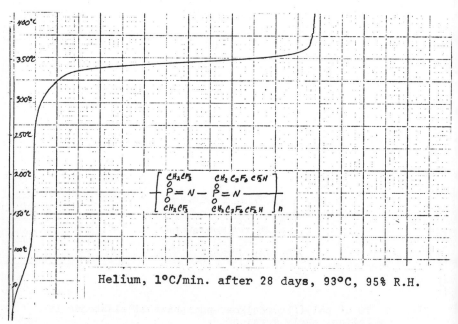

Helium, 1°C/min. after 28 days, 93°C, 95% R.H.

Fig. 12   TG of poly(fluoroalkoxy-phosphazene) elastomer in
helium, early version after humid exposure

148

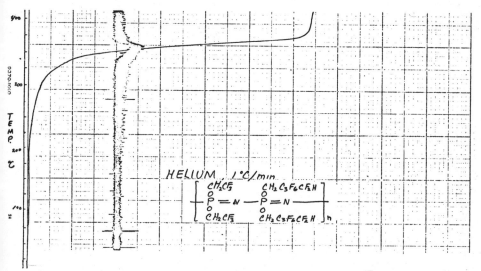

Fig. 13 TG of poly(fluoroalkoxy-phosphazene) elastomer in
helium, improved version

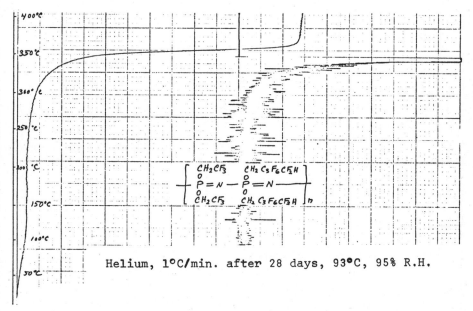

Fig. 14 TG of poly(fluoroalkoxy-phosphazene) elastomer in
helium, improved version after humid exposure

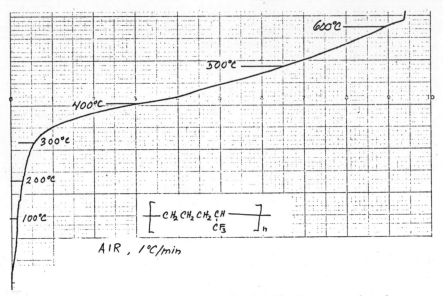

Fig. 15 TG of poly(ethylene-co-propylene) elastomer in air

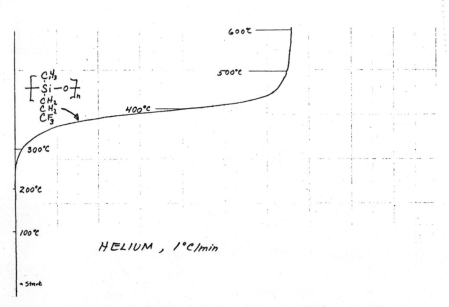

Fig. 16 TG of poly(trifluoropropylmethyl-siloxane) elastomer in helium

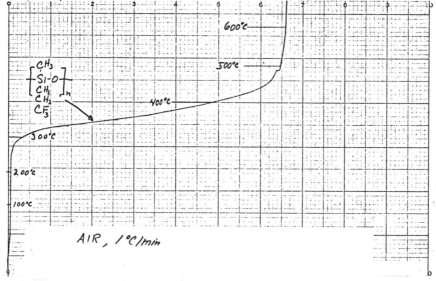

Fig. 17 TG of poly(trifluoropropylmethyl siloxane) elastomer
in air

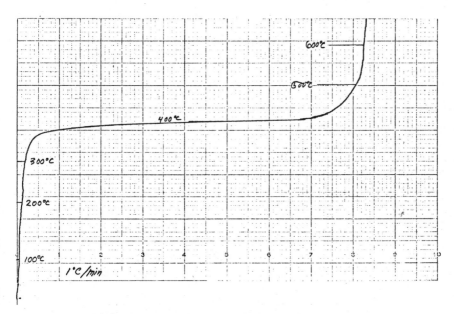

Fig. 18 TG of poly(trifluoropropylmethyl siloxane) fuel tank
sealant in air

151

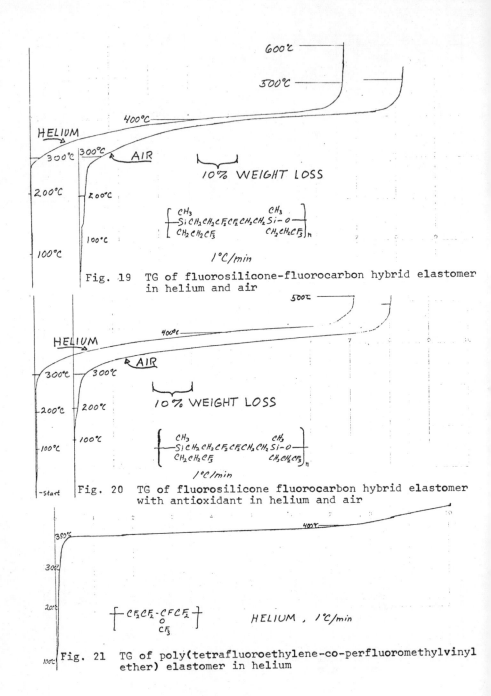

Fig. 19 TG of fluorosilicone-fluorocarbon hybrid elastomer
in helium and air

Fig. 20 TG of fluorosilicone fluorocarbon hybrid elastomer
with antioxidant in helium and air

Fig. 21 TG of poly(tetrafluoroethylene-co-perfluoromethylvinyl
ether) elastomer in helium

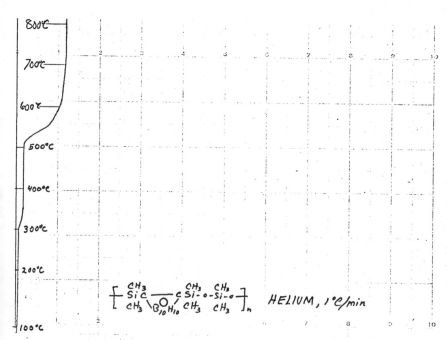

Fig. 22 TG of dimethylsiloxane-m-carborane elastomer in helium

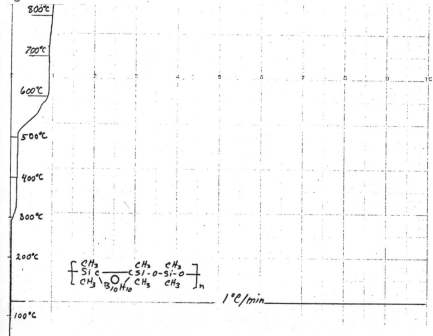

Fig. 23 TG of dimethylsiloxane-m-carborane elastomer in air

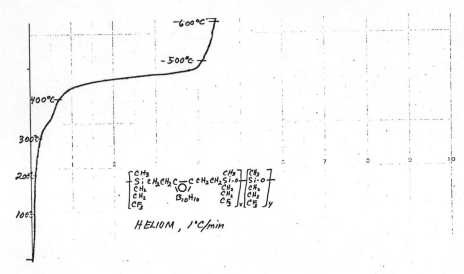

Fig. 24   TG of trifluoropropylmethyl siloxane-m-carborane
elastomer in helium

NON-ISOTHERMAL KINETICS

T. Ozawa

Electrotechnical Laboratory
Tanashi, Tokyo, Japan

ABSTRACT

The relations for non-isothermal kinetics of diffusion, nucleation-and-growth process and chemical reaction consisting of a single unit process are derived. From these relations, methods for kinetic analysis of thermoanalytical data are proposed. The non-isothermal kinetic equation of chemical reaction is applied to non-isothermal deterioration of electrical insulating materials, and its applicability is demonstrated. The methods for kinetic analysis of thermoanalytical data are applied to diffusion of impurities from electronic packaging resin and crystallization of polymeric materials. Finally, results of critical investigation of methods for kinetic analysis of thermoanalytical data proposed by some workers are described, and the present status of non-isothermal kinetics is also mentioned.

INTRODUCTION

Non-isothermal kinetics was first proposed by Flynn [1]. However, as Flynn pointed out, the concept had already been set forth many years before. For example, in 1928 Akahira proposed a new concept "equivalent temperature" at which the chemical reaction proceeds isothermally to the same extent as the chemical reaction proceeds non-isothermally [2], and furthermore, he published a table of exponential integral function necessary to calculate the extent of chemical reaction proceeding at a constant rate of heating or cooling [3]. The non-isothermal kinetics was not developed at that time, presumably because there was not great necessity for the non-isothermal kinetics. In 1958, Freeman and Carroll published their method for kinetic analysis of thermoanalytical data [4]. Since then, various and numerous methods for kinetic analysis have been proposed [5]. This tendency stimulated new development of the non-isothermal kinetics. Some methods for the kinetic analysis can be derived without the non-isothermal kinetics, but sound and widely applicable methods are derived on the base of the non-isothermal kinetics. For instance, when one analyses thermoanalytical curves of parallel reaction by some methods, false and unreal kinetic parameters are obtained, as shown later. Thus, various types of isothermal kinetic equations should be extend to non-isothermal relations.

In this paper, the non-isothermal kinetics of diffusion, chemical reaction consisting of a single unit process and the process of nucleation-and-growth are described. From the

155

non-isothermal kinetic equations, methods for the kinetic analysis are derived. These equations and methods are applied to a few processes occuring in some electrical insulating materials and polymeric materials.

KINETICS OF CHEMICAL REACTIONS

Generally, the fundamental kinetic equation of chemical reaction is as follows;

$$\frac{dx}{dt} = kg(x),$$ (1)

where x, t, k and $g(x)$ are the amount of reacting chemical species, the time, the rate constant and a function of x representing the mechanism of chemical reaction, respectively, and usually

$$k = A\exp(-\frac{\Delta E}{RT}),$$ (2)

where A, ΔE, R and T are the pre-exponential factor, the activation energy, the gas constant and the absolute temperature respectively.

The fundamental equation (1) should be solved in the non-isothermal condition, in which the rate constant is no longer the constant but a function of the temperature such as equation (2). Then

$$\int_{x_0}^{x} \frac{dx}{g(x)} = A\int_{t_0}^{t} \exp(-\frac{\Delta E}{RT})dt,$$ (3)

where x_0 is the amount of x at $t=t_0$. Introducing a new concept reduced time, θ, which equals to $\int \exp(-\Delta E/RT)dt$ [6], we get

$$G(x) - G(x_0) = A\theta - A\theta_0,$$ (4)

where $G(x)$ is equal to $\int dx/g(x)$ and θ_0 is the reduced time at $t=t_0$. Further [7],

$$\frac{dx}{dA\theta} = g(x),$$ (5)

and

$$\frac{dx}{dA\theta} = \frac{dx}{dt} / A\exp(-\frac{\Delta E}{RT}).$$ (6)

In order to derive the relation between x or dx/dt and t by using these equations, the relation between t and θ should first be found for various types of temperature change. Because the derived relations are described elsewhere [6], the relation only for the case in which the temperature chagnes at a constant rate of heating or cooling, a, between T_1 and T_2 ($T_1>T_2$) is shown below;

$$\theta = \frac{\Delta E}{aR}\{p(\frac{\Delta E}{RT_1}) - p(\frac{\Delta E}{RT_2})\},$$ (7)

156

where p is the p-function proposed by Doyle [8].

In some cases, we measure the quantity proportional to x or dx/dt directly in thermal analysis, but in other cases, the recorded property in thermal analysis is not proportional to x or dx/dt. For instance, in differential scanning calorimetry (DSC), the height of DSC curve from its base line is proportional to dx/dt, if the changes of heat capacity due to both the reaction and the temperature change are neglisible, and if the change of the heat of reaction with increase of the temperature is neglisibly small. However, when we observe random scission in main chain of polymer by thermogravimetry (TG), the weight change is not proportional to x, $i. e.$, the fraction of broken bond in the main chain of polymer, but a complicated function of x as shown below, although the weight change is directly proportional to the amount of product or reacting species in other cases (for instance, thermal decomposition of some carbonates). For the random scission in the main chain of a polymer,

$$1 - C = (1 - x)^{L-1}\{1 + x\frac{(N - L)(L - 1)}{N}\}, \tag{8}$$

where C, N and L are the conversion, the initial degree of polymerization and the least degree of polymerization of unvolatilized polymer, and x increases by the first order reaction. Thus, we should introduce a function representing the relation between the conversion of the measured quantity and the amount of reacting species, x,

$$C = f(x). \tag{9}$$

Combining the above-mentioned equations, we have

$$C = \Phi(A\theta) \tag{10}$$

and

$$\frac{dC}{dA\theta} = \frac{df(x)}{dx} \frac{dx}{dt} / A\exp(-\frac{\Delta E}{RT}). \tag{11}$$

The concrete forms of the equations for a few types of chemical reactions and their plots are shown elsewhere [7].

It is inferred from the above equations that an experimental master curve can be drawn when the experimental data of the conversion or the rate of conversion versus the actual time observed both isothermally and non-isothermally is transformed to the relation between the conversion or the reduced rate of conversion ($dC/dA\theta$) and the reduced time. Because the reduced time is proportional to the actual time in the isothermal run, the transformation of the non-isothermal data to the experimental master curve corresponds to the conversion of the non-isothermal data to the isothermal data at infinitely high temperature. The reduced time is a measure of the extent of chemical reaction both in the isothermal run and in the non-isothermal run. When the reduced times are equal to each other in two experiments of the same reaction, the extent of the reaction is the same in each, and the conversion is also the same.

The above-mentioned relations hold only if the reaction

consists of a single process and if the observed property is dependent only on the amount of reacting species and independent of the temperature and the experimental time scale (e. g., the frequency of the external field). These two restrictive conditions should be borne in mind. Some reactions consist of multiple processes, but the overall kinetic equation of the reactions is the same as the equation (1). An example is depolymerization of the polymers in which radical is formed by chain scission and monomer is unzipped from the radical end. The radical concentration diminishes by the combination of the two radicals. On the assumption that the rate of the radical formation is equal to the rate of the radical annihilation, the rate of the monomer unzipping, i. e., the rate of the weight loss is proportional to the residual weight, so that the overall kinetic equation is the same as the equation (1). Strictly speaking, the non-isothermal kinetic equations mentioned above can not be applied to the depolymerization, because the assumption of the constant concentration of the radical in the steady state no longer holds, but if the concentration of the radical in the non-isothermal run is almost equal to the steady state concentration of the radical in the isothermal run at that temperature, in other words, if the time elapsed from the formation of one particular radical to its annihilation is small enough in comparison to the rate of temperature change and the change of the steady state radical concentration due to the temperature change, the equations may be applicable to the non-isothermal depolymerization. This author and his co-worker have applied the methods based on the above-mentioned non-iso-thermal kinetic equations to the thermoanalytical data of depolymerization of polymethyl methacrylate; reasonable kinetic parameters were obtained [9].

Another example is diffusion-controlled processes. In the steady state of an isothermal experiment, the conversion is proportional to the square root of the time. When the overall reaction is diffusion-controlled in the experimental temperature range, the conversion may be approximately proportional to the square root of the reduced time.

Now, an example of application of the non-isothermal kinetic relation is shown. Electrical insulating material deteriorates mainly due to thermo-oxidative reaction, so that the reduced time is a measure of thermal deterioration of the characteristic of the materials [6]. In some electrical motors such as traction motors, the insulating materials deteriorate non-iso-thermally. In Table I, the reduced time of linear heating and cooling in the indicated temperature range is compared with the isothermal reduced time at the mean temperature. The non-iso-thermal reduced time is larger than the isothermal one, because the decrease of the rate in the temperature range lower than the mean temperature is smaller than the increase of the rate in the higher temperature range.

For this reason, Akahira [2] proposed a new concept, the equivalent temperature, T_e, at which the isothermal reduced time is equal to the non-isothermal reduced time, i. e.,

$$t \exp\left(-\frac{\Delta E}{R T_e}\right) = \int_0^t \exp\left(-\frac{\Delta E}{RT}\right) dt. \tag{12}$$

158

The comparison between non-isothermal deterioration and the isothermal deterioration is shown in Figs. 1 and 2 [10]. In Fig. 1, the deterioration of varnish glass cloth of alkyd resin is compared. The activation energy of the deterioration is about 22 kcal/mol and 210°C is the equivalent temperature corresponding to the linear cyclic heating and cooling between 155°C and 240°C. The dielectric breakdown voltage measured with flat electrodes is in very good agreement. The weight loss of polyvinylformal-emanelled wire is shown in Fig. 2. The activation energy, equivalent temperature and the upper and lower limits of the cycle are the same as those of the varnish glass cloth. The agreement is also very good. However, the decrease of dielectric breakdown voltage of the varnish glass cloth measured with curved electrodes under 2% elongation for the non-isothermal deterioration is larger than that for the isothermal deterioration. As to withstand voltage test of twisted pair of the enamelled wire, the situation is the same; the deterioration is faster in the non-isothermal deterioration than in the isothermal deterioration. These two characteristics may depend on the texture as well as the chemical structure, and they may reflect the change of the texture as well.

DIFFUSION

The fundamental equation of diffusion is

$$D\nabla^2 \xi = \partial\xi/\partial t, \tag{13}$$

where D, ∇^2 and ξ are the diffusion constant, a Laplacian operator and the concentration of diffusing species, respectively. The temperature dependence of the diffusion constant may be expressed in the next equation,

$$D = D_0 h(T). \tag{14}$$

Introducing the reduced time, θ, which is equal to $\int h(T)dt$, the generalized fundamental equation of diffusion is obtained [11],

$$D_0\nabla^2 \xi = \partial\xi/\partial\theta. \tag{15}$$

This equation is the fundamental equation applicable both to the isothermal case and to the non-isothermal case.

Solving the equation (15) on certain initial and boundary conditions, we can obtain the change of the concentration of the diffusing species. Because the equation (15) is the same in its form as the fundamental isothermal equation (13), the typical solutions are easily found [12]. The following equation is a typical solution for an infinite plate, and the relations between the conversion or the reduced rate of conversion of volatilization and the reduced time are given on the initial condition of unifrm concentration within the specimen and on the boundary condition of zero concentration outside of the specimen:

$$C = 1 - 8 \sum_{i=0}^{\infty} \exp\{-(2i+1)^2\pi^2\theta\}/(2i+1)^2\pi^2 \tag{16}$$

159

$$\frac{dC}{d\theta} = 8 \sum_{i=0}^{\infty} \exp\{-(2i+1)^2 \pi^2 \theta\}, \tag{17}$$

where θ is $D_0\theta/r^2$ and r is the thickness of the plate, and

$$\frac{dC}{d\theta} = \frac{dC}{dt}/(D/r^2). \tag{18}$$

The relation between the rate of conversion and θ is plotted in Fig. 3.

As is clearly seen, the conversion and the reduced rate of conversion are also both single-valued functions of θ or θ. This situation is the same as that of chemical reaction. By utilizing this similarity, the method for kinetic analysis of thermoanalytical data which is applicable both to the chemical reaction and the diffusion can be derived.

METHODS FOR KINETIC ANALYSIS OF THERMOANALYTICAL DATA

Two methods for kinetic analysis of thermoanalytical data are devised from the above-mentioned kinetic equations. One is applicable to the case that the temperature dependence of the rate constant or the diffusion constant is Arrhenius type. The other is applicable to the case of general temperature dependence, but the mechanism must be first assumed. Both are applicable to the chemical reaction as well as to the diffusion.

Because the conversion is a single-valued function of the reduced time both in the non-isothermal kinetics of chemical reaction and in that of diffusion, the reduced time is the same at the same conversion, $i.\ e.$, when the observed process proceeds to a given conversion at T_1 for the heating rate a_1, and at T_2 for a_2, the next equation holds [6, 11],

$$p\left(\frac{\Delta E}{RT_1}\right)/a_1 = p\left(\frac{\Delta E}{RT_2}\right)/a_2, \tag{19}$$

because the temperature is raised usually from a low temperature at which the process scarcely proceeds. [$\theta_0 = 0$ in the equation (4).] The p-function can be approximated as follows [8];

$$p(y) = \exp(-y)/y^2 \tag{20}$$

or

$$\log p(y) = -2.315 - 0.4567y. \tag{21}$$

From the equation (19), we have

$$\log a_1/T_1^2 + \Delta E/2.303RT_1 = \log a_2/T_2^2 + \Delta E/2.303RT_2. \tag{22}$$

or

$$\log a_1 + 0.4567\Delta E/RT_1 = \log a_2 + 0.4567\Delta E/RT_2 \tag{23}$$

At the peak of the curve of dC/dt versus t, the following equation is obtained. Namely,

$$C = \Phi(\theta) \tag{24}$$

where Θ equals to $A\theta$ for the chemical reaction. Differentiating the above equation twice, we have

$$\frac{d^2 C}{dt^2} = \frac{d^2 \Phi(\Theta)}{d\Theta^2} A^2 \exp(-\frac{2\Delta E}{RT}) + \frac{\alpha A \Delta E}{RT^2} \frac{d\Phi(\Theta)}{d\Theta} \exp(-\frac{\Delta E}{RT}).$$ (25)

Utilizing the approximate equation (20) and equating to zero,

$$\frac{d^2 \Phi(\Theta_m)}{d\Theta_m^2} \Theta_m + \frac{d\Phi(\Theta_m)}{d\Theta_m} = 0,$$ (26)

where the subscript m denotes the peak. Because this equation contains only Θ_m and the constant, Θ_m does not depend on the heating rate, and it is constant, so that the equations (22) and (23) hold also at the peak.

The plot $\log a$ or $\log a/T^2$ versus $1/T$ at a given conversion or at the peak gives a straight line and the activation energy can be estimated from the slope. In some cases, the estimated activation energy changes with increase of the conversion. A few examples were described by David [13] for a few electrical insulating materials and by Popa and his co-worker [14] for thermo-oxidative degradation of polyvinylalcohol, and another example shown later is obtained by analysing the theoretically calculated thermoanalytical curves for a parallel and competitive reaction. The change of the activation energy seems to be due to the fact that the process does not consist of a single unit process; this means that the method and the non-isothermal kinetic equations mentioned above can not be applied to the case, to which any methods proposed till now can not be applied.

Examples of the plot of the logarithm of heating rate versus the reciprocal absolute temperature at the peak are described elsewhere [9, 15], as well as the plots of the logarithm of heating rate versus the reciprocal absolute temperature at a given conversion [6]. An application of this plot to thermoanalytical data other than TG, EGA and DSC is shown by Kambe [16]. By using the thus estimated value of the activation energy, the experimental data can be transformed to the relation beween C or $dC/d\theta$ and θ. This is an experimental master curve, and it is then compared with the theoretical relation between C or $dC/d\theta$ and θ. By this comparison, the mechanism can be elucidated and the pre-exponential factor can also be estimated. This procedure has been applied to thermal depolymerization of polymethyl methacrylate [9], thermal degradation of nylon 6 [6] and volatilization of hydrogen chloride from epoxy resin [15].

The other method is to estimate the rate constant or the diffusion constant directly. The value of $dC/d\theta$ at a given conversion is known for a certain mechanism [see the equations (5), (16) and (17).] On the other hand, $dC/d\theta$ is in relation with dC/dt and the rate constant or the diffusion constant divided by the square of r, as is seen in the equations (11) and (18). We can estimate the rate constant or the diffusion constant in the following way. First, the conversion is estimated by using the data of dC/dt versus t. The $dC/d\theta$ is calculated at the estimated conversion for an assumed mechanism. By comparison of $dC/d\theta$ with dC/dt, the rate constant or the diffusion constant is estimated along the conversion or the temperature. If the assumed mechanism is correct, the rate constants or the diffusion constants estimated at different heating rates are superposed on each other.

Examples obtained by the second method can be seen for the thermal depolymerization of polymethyl methacrylate [7] and the volatilization of hydrogen chloride and toluene from epoxy resin [15, 11].

PROCESS OF NUCLEATION AND GROWTH

The isothermal kinetic equation of the process initiated by nucleation and proceeding by growth of the nuclei, such as crystallization and some solid reactions, is well known as Avrami-Erofeev equation [17-21];

$$-\ln(1 - C) = Zt^n \tag{27}$$

where Z is a constant and the value of n is listed in Table II.

In order to extend the equation (27) to a non-isothermal equation, the logic to derive the equation (27) by Evans [19] is very useful [22], and it is the way of analogy. We shall follow Evans' way and extend each step to the non-isothermal one. Evans compared a nucleus to a raindrop falling upon a pond and its growth to expansion of a circular wave caused by the raindrop. While the wave can pass over a particular point many times, the front of growth of the new phase can pass over the point only once. This is one difference between the expansion of the wave caused by the raindrop and the process of nucleation and its growth, and the other is the dimension of growth. In order to remove the former difference, Evans first calculated the expectancy of the number of waves passing over the point, and he estimated the fraction of the pond where any wave has not yet passed over by using Poisson's distribution function.

Let us first calculate the expectancy for the crystallization at the constant rate, a, of cooling from the melting point. The distance γ which the front of growth traverses from the time t_1 to t is given by

$$\gamma = \int_{t_1}^{t} v(T)\,dt, \tag{28}$$

where $v(T)$ is the linear growth rate as a function of the temperature, T, and integration gives

$$\gamma = \frac{1}{a}\{R_c(T) - R_c(\tau)\}, \tag{29}$$

where $R_c(T)$ equals to $\int_{T_m}^{T} v(T)\,dT$, and T, τ and T_m are the temperature at t, that at t_1 and the melting point above which the process cannot proceed. The number j of the nuclei per unit volume or area which form from T_m to τ is given by

$$j = \int_{T_m}^{\tau} \nu(T)\,dt \tag{30}$$

$$= \frac{1}{a}\int_{T_m}^{\tau} \nu(T)\,dT \tag{31}$$

$$= N_c(\tau)/a, \tag{32}$$

where $\nu(T)$ is the nucleation rate as a function of T and $N_c(\tau)$ is equal to $\int_{T_m}^{\tau} \nu(T)\,dT$.

The expectancy, $E(T)$, that the number of the waves pass over the particular point is then as follows;

$$E(T) = g\int_{T_m}^{T} j\gamma^m d\gamma \tag{33}$$

$$= (g/a^{m+2})\int_{T_m}^{T} N_c(\tau)\{R_c(T) - R_c(\tau)\}^m v.(\tau)\,d\tau, \tag{34}$$

where $m+2$ equals to n, and n and g are listed in Table II. As the integration of the equation (34) depends only on T,

$$E(T) = \chi_c(T)/a^n, \tag{35}$$

where

$$\chi_c(T) = g\int_{T_m}^{T} N_c(\tau)\{R_c(T) - R_c(\tau)\}^m v(\tau)\,d\tau. \tag{36}$$

This function should be called the cooling function of the process.

Thus, the conversion, C, at temperature, T, and cooling rate, a, is given by

$$1 - C(T) = \exp\{-\chi_c(T)/a^n\} \tag{37}$$

because the probability that the k waves passing over the particular point equals to $\exp(-E)E^k/k!$. When the substance is heated at a rate, a, from a temperature at which the process scarecely occurs, the following similar equation holds,

$$1 - C(T) = \exp\{-\chi_h(T)/a^n\}, \tag{38}$$

where the lower limit of the temperature of the integration in the equations (29), (31), (34) and (36) is zero.

The equations for the rate of the process are given as follows;

$$dC/dt = \{1 - C(T)\}\chi_c'(T)/a^{n-1} \tag{39}$$

$$dC/dt = \{1 - C(T)\}\chi_h'(T)/a^{n-1}, \tag{40}$$

where $\chi_c'(T)$ and $\chi_h'(T)$ are the temperature derivatives of the original functions.

If pre-determined nuclei exist before cooling, the number of nuclei becomes constant and equals to the density ω of the nuclei. Then, on cooling,

$$1 - C(T) = \chi_c(T)/a^n, \tag{41}$$

where n is listed in Table II, and

163

$$\chi_c(T) = g\omega\int_{T_m}^{T} \{R_c(T) - R_c(\tau)\}^m v(\tau)\,d\tau. \tag{42}$$

If we now observe the process at different rates of cooling or heating, the kinetic parameters mentioned above can be obtained,

$$\log[-\ln\{1 - C(T)\}] = \log\chi(T) - n\log a. \tag{43}$$

If we plot $\log[-\ln\{1 - C(T)\}]$ against $\log a$ at a given temperature, a straight line can be obtained, and n and $\chi(T)$ can be estimated from the slope and the intercept of the line. $\chi(T)$ is obtained as a function of the temperature from the plots at different temperatures. For the case of $n=1$, $R(T)$ or $v(T)$ is obtained from $\chi(T)$ or $\chi'(T)$.

The method for kinetic analysis is applied to DSC curves of crystallization of polyethylene terephthalate [22] and polytetrafluoroethylene [23]. Fig. 4 shows curves of the conversion versus the temperature for the crystallization of polytetrafluoroethylene obtained by integration of DSC curves, and Fig. 5 is the plot of $\log\{-\ln(1 - C)\}$ versus $\log a$. The slopes of the lines in Fig. 5 are approximately unity; the facts mean that the crystallite of polytetrafluoroethylene grows one-dimensionally from a pre-determined nucleus.

CRITICAL INVESTIGATION OF THE METHODS FOR KINETIC ANALYSIS OF THERMOANALYTICAL DATA

In the course of deriving the non-isothermal kinetic equations and the methods for kinetic analysis, this author avoided using particular and concrete kinetic formulae and tried to use general and abstract kinetic relations. The purpose is to derive the generally applicable non-isothermal kinetic relations and the generally applicable methods for kinetic analysis. Although numerous methods for kinetic analysis of thermoanalytical data have been proposed till now [5], the most of them are based on a particular kinetic formula. The dangerous tendency of these methods is demonstrated [24]. A few examples are shown below.

In Fig. 6 are shown the result of kinetic analysis of the theoretically calculated TG (or EGA) curves of the random scission in the main chain of polymer by the method of Freeman and Carroll [4]. Even though the method can not be applied to the case, nearly straight lines are obtained, and the activation energy and the order are estimated to be about 70 kcal/mol and 1.3, respectively, while the true values are 40 kcal/mol, 4 and 10^{12}/sec for the activation energy, the least degree of polymerization of unvolatilized polymer and the pre-exponential factor, respectively. The method of Coats and Redfern was also applied to the curves, and the plots on the assumption of the first order reaction are shown in Fig. 7. The unreal kinetic parameters are also estimated; the activation energy is about 64kcal/mol and the logarithm of the pre-exponential factor is 21.3 (1/sec). The first method described in this article was applied, and the activation energy and the logarithm of the pre-exponential factor are estimated to be 40.2 kcal/mol and 12.22 (1/sec), respectively. The mechanism is elucidated to be the

164

random scission in the main chain of polymer with the least degree of polymerization of unvolatilized polymer of 3.

These methods were also applied to the theoretical curves of parallel and competitive reaction consisting of two unit processes [23], both of which are the first order reaction; the activation energy and the pre-exponential factor of one reaction are 40 kcal/mol and 2×10^{11}/sec, and those of the other reaction are 60 kcal/mol and 10^{19}/sec, respectively. The theoretically calculated curves at different heating rates are shown in Fig. 8. The results of kinetic analysis are shown in Figs. 9, 10 and 11. Although all methods can not be applied, nearly straight lines can be drawn. Incorrect kinetic parameters are obtained by the methods of Freeman and Carroll and that of Coats and Redfern. On the other hand the apparent activation energy is dependent on the converion by this author's method; the fact suggests that the method can not be applied.

The demonstration described above shows us the possibility of obtaining unreal and false kinetic parameters by the methods based on the limited fundamental equations and the necessity of estimating the kinetic parameters by utilizing data obtained at different heating rates.

UNSOLVED PROBLEMS OF NON-ISOTHERMAL KINETICS

In this article, two ways of extending the isothermal kinetic equations to the non-isothermal kinetic equations are described; one is the use of the reduced time and the generalization of the isothermal equations to the non-isothermal equations, and the other is the logic of expectancy and Poisson's distribution function. The non-isothermal kinetic equation of the n-th order of reaction can also be derived by the second way. These ways are very useful. However, there remain many unsolved kinetic mechanisms for which the non-isothermal kinetic equations have not yet been derived. The situation is shown schematically in Fig. 12. For competitive and/or consecutive reactions, introduction of multiple reduced times is necessary, because each unit process has its own reduced time. Hence, the system of multidimensional reduced times is necessary to express them, and difficulty of solving it is in the fact that the relation among the reduced times is very complicated. On the other hand, effectiveness of the second logic is restricted within the linear heating or cooling.

165

REFERECES

1. J. H. Flynn, "Thermal Analysis", R. F. Schenker, Jr., and P. D. Garn, Ed., Academic Press, New York, Vol. 2, p. 1111.
2. T. Akahira, Sci. Papaers Inst. Phys. Chem. Res., 9, 165 (1928).
3. T. Akahira, *ibid.*, Table No. 3, 181 (1929).
4. E. S. Freeman and B. Carroll, J. Phys. Chem., 62, 394 (1958).
5. For example, J. H. Flynn and L. A. Wall, J. Res. Natl. Bur. Stds., 70A, 487 (1966); J. Sestak, V. Satava and W. W. Wendlandt, Thermochim. Acta, 7, 333 (1973).
6. T. Ozawa, Bull. Chem. Soc. Japan, 38, 1881 (1965).
7. T. Ozawa, J. Thermal Analysis, 2, 301 (1970).
8. C. D. Doyle, J. Appl. Polymer Sci., 6, 639 (1962); Nature, 207, 290 (1965).
9. R. Sakamoto, T. Ozawa and M. Kanazashi, Thermochim. Acta, 3, 291 (1972).
10. T. Ozawa, T. Kaneko and Y. Takahashi, Preprint Annual Meeting Inst. Elect. Engrs. Japan (Tokyo, 1974).
11. T. Ozawa, J. Thermal Analysis, 5, 563 (1973); *ibid.*, 6, No. 3, to be published.
12. J. Crank, The Mathematics of Diffusion, Oxford University Press, London, 1957.
13. D. J. David, Insulation, 13, No. 12, 38 (1967).
14. M. Popa, C. Vasile and I. A. Schneider, J. Polymer Sci., Polymer Chem. Ed., 10, 3679 (1972).
15. R. Sakamoto, Y. Takahashi and T. Ozawa, J. Appl. Polymer Sci. 16, 1047 (1972).
16. H. Kambe, M. Kochi, T. Kato and M. Murakami, This Proceedings.
17. M. Avrami, J. Chem. Phys., 7, 1103 (1939); *ibid.*, 8, 212 (1940); *ibid.*, 9, 177 (1941).
18. F. von Goler and G. Sachs, Z. Physik., 77, 281 (1932).
19. W. A. Johnson and R. F. Mehl, Trans. Am. Inst. Mining Met. Engrs., 135, 416 (1939).
20. U. R. Evans, Trans. Faraday Soc., 41, 365 (1945).
21. B. V. Erofeev, Dokl. Akad. Nauk USSR, 52, 511 (1946).
22. T. Ozawa, Polymer, 12, 150 (1971).
23. T. Ozawa, unpublished data.
24. T. Ozawa, J. Thermal Analysis, to be published.

Table I. The comparison of the non-isothermal reduced time with the isothermal reduced time at the mean temperature.

The activation energy (kcal/mol)	Linear heating and cooling*	Isothermal at the mean temperature*
20	1	0.83
25	1	0.75
30	1	0.67
35	1	0.58
40	1	0.50

*The temperature is changed linearly between 130 and 170°C, and the mean temperature is 150°C.

Table II. The relation between the mechanism and
the values of n and the geometrical factor, g.

Nucleation	Dimension of growth	n	g
	one-dimensional	2	area of nucleus
random	two-dimensional	3	2π
	three-dimensional	4	π
	one-dimensional	1	area of nucleus
pre-determined	two-dimensional	2	2π
	three-dimensional	3	π

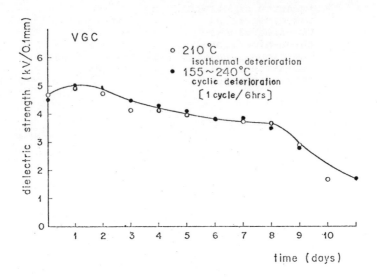

Fig. 1. The change of dielectric strength of the varnish glass cloth of alkyd resin.

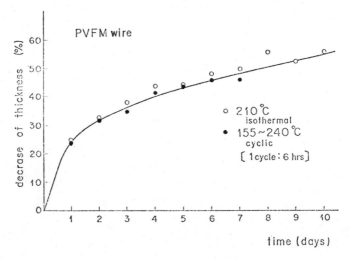

Fig. 2. The weight loss of the polyvinylformal-enamelled wire.

168

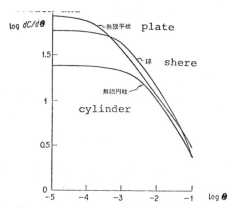

Fig. 3. The relations between dC/dΘ and Θ for diffusion on the initial condition of uniform concentration within the specimen and on the boundary condition of zero concentration outside ot the specimen.

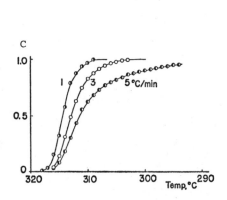

Fig. 4. The conversion of crystallization of poly-tetrafluoroethylene at different cooling rates indicated in the figure.

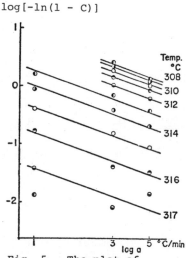

Fig. 5. The plot of log -ln(1-C) versus log a for the cry-stallization of poly-tetrafluoroethylene.

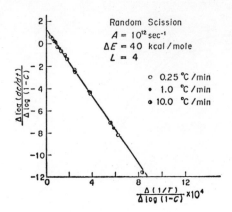

Fig. 6. The plots for kinetic analysis of the theoretically calculated TF (or EGA) curves of the random scission in the main chain of polymer by the method of Freeman and Carroll. The heating rate is indicated in the figure.

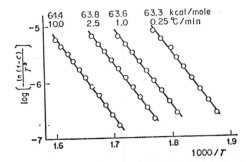

Fig. 7. The plots for the kinetic analysis of the same curves as in Fig. 6 by the method of Coats and Redfern. The first order reaction is assumed. The heating rate is indicated in the figure.

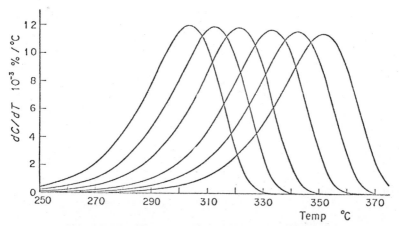

Fig. 8. The theoretically calculated thermoanalytical curves
of the parallel and competitive reaction consisting
of two first order reactions.   The heating rates
are 0.25, 0.5, 1.0 2.5, 5.0 and 10.0°C/min from
left to right.

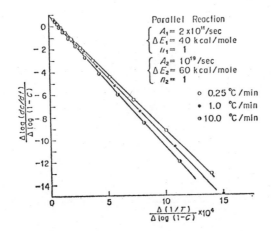

Fig. 9. The plots for the kinetic analysis of the curves in
Fig. 8 by the method of Freeman and Carroll.   The
heating rate is indicated in the figure.

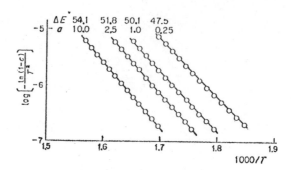

Fig. 10.   The plots for the kinetic analysis of the curves in
            Fig. 8 by the method of Coats and Redfern on the
            assumption of the first order reaction.   The
            estimated activation energy and the heating rate
            are indicated in the figure.

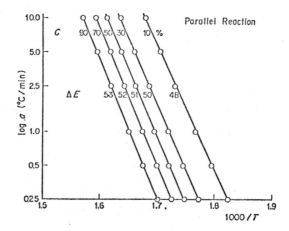

Fig. 11.   The plots of the logarithm of heating rate against
            the reciprocal absolute temperature at the indicated
            conversion (the author's method).   The estimated
            apparent activation energy written in the figure
            is dependent on the heating rate.

172

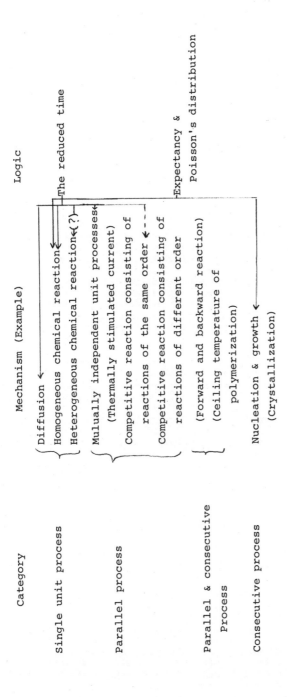

Fig. 12.   The present status of the non-isothermal kinetics.   The arrow means the applicability.

173

THERMODYNAMICS OF MESOPHASE TRANSITIONS

Edward M. Barrall II

IBM Research Laboratory
San Jose, California

The melting of a crystalline substance is a familiar phase transition
The highly structured solid melts to the isotropic liquid phase at a
well-defined temperature and with a characteristic heat of fusion.

In contrast, a considerable number of organic compounds--more than
3000--do not melt directly from crystalline solid to isotropic liquid.[1,2]
Instead, the substance passes through an intermediate phase, a mesophase.
In such a case, two phase transitions are involved: at a lower temperature,
a transition from crystalline solid to mesophase, and at a higher temperature,
a transition from mesophase to isotropic liquid.

The mesophase appears as a turbid liquid, and the transitions can be
observed visually or by other common techniques. The term liquid crystal
has been used interchangeably with mesophase. To a lesser extent, this state
has been referred to as anisotropic or paracrystalline.[1]

Mesophases are ordered on a molecular level, yet possess some of the
mechanical properties of liquids. The ordered domains in mesophases contain
typically about 10^5 molecules. However, the sense of arrangement of the
molecules is not the same in all mesophases. Three basic types are
recognized: nematic, smectic, and cholesteric mesophases. A single compound
can exhibit more than one mesophase.

Figure 1 illustrates the general structural features of the three
mesophases. The nematic type is the simplest. In pure compounds, e.g.,
p-azoxyanisole, the nematic mesophase is thought to consist of bundles of
rodlike structures. The only structural restriction in nematic mesophases
is that the molecules are parallel or nearly parallel within a bundle.

In the smectic mesophase, molecules are arranged side by side in a series
of stratified layers,[2] e.g., diethyl p,p'-azoxydibenzoate. Molecules in the
layers may be arranged in a regular or random side-by-side spacing. The long
axes of the molecules are parallel to one another and essentially
perpendicular to the base plane of the layer. The layer can be one or more
molecules thick. In general, the smectic mesophase has greater order than
the nematic.

The third type is the cholesteric mesophase; it is formed principally by
derivatives of cholesterol, e.g., cholesteryl benzoate, but not by cholesterol
itself. The cholesteric mesophase resembles the smectic mesophase in that
molecules are arranged in layers. Within each layer, however, the parallel

arrangement of molecules is reminiscent of the nematic mesophase. The cholesteric layers are thin with the long axis of molecules parallel to the plane of the layers. A screw axis runs through the layers.[2]

A relatively new but widely applied focus in mesophase research is the calorimetric study of the phase transitions. From the molar heats of transition, q, the entropy change of transition is calculated from the familiar relationship, $\Delta S = q/T$, where T is the absolute temperature. Only about a dozen mesophase-forming compounds had been subjected to such calorimetric study prior to 1960. Indeed, Brown and Shaw in their exhaustive review of the literature to 1957 could find heat of fusion data on only four materials.[1] Thermodynamic data are now known on entire homologous series of mesophase forming materials. Well over 600 materials have relatively complete temperature and entropic data available.

A knowledge of both the temperature and the heat of transition is necessary if the principals of physical analysis are to be applied to mesophase forming systems. From this information the transition entropy may be calculated. This acts as the key for evaluating the type and degree of order present in many systems.

In the past it was customary to measure and record only the temperature of mesophase transitions. While this has been very profitable in a number of systems,[3] temperature alone cannot furnish a sufficiently encompassing basis for prediction of mesophase behavior. Most attempts to predict mesophase transition temperatures, even in homologous series, using only a knowledge of the transition temperatures of other members have not met with a high degree of success. A survey of the transition temperatures of any homologous series indicates important trends. However, "exceptions" appear. A list of transition entropies is a somewhat more regular indication of important thermal trends.

The recent proliferation of thermodynamic information is principally due to two causes: pressing need for the information and great improvements in thermal instrumentation. Liquid crystals are currently receiving attention in such varied areas as computer display and biomedical work.[4] Liquid crystalline materials when placed between conductors or in a strong magnetic field undergo dramatic changes in optical rotation and/or light scattering properties. It is this set of properties that have interested computer display developers and video display designers. The search for homologues with room temperature liquid crystal ranges is a driving force in the large scale study and synthesis of a broad range of materials. Every known mesophase type has been employed in some type of display. Progress has been so rapid that in only three years commercial calculator, wristwatch and thermometric displays have appeared on the open market. The low voltage and minute current requirements make these displays compatible with the most sophisticated solid state devices. It is projected that in ten years liquid crystal displays will be competing closely with cathode ray and glow type tubes for the indicator and television markets. Large industrial and governmental research departments are actively engaged in a wider range of liquid crystal research than would have been thought possible a few years ago. Not since the active days of Lehmann and Vorlander have so many mesophase materials been made. The question of thermodynamic properties naturally arises in these research efforts. Recent application of the methods of differential thermal analysis (DTA) and differential scanning calorimetry (DSC) has greatly facilitated the determination of temperatures, heats of transition and heat capacity of various phases. The approach of classical calorimetry has not been neglected. At least one laboratory[5] has made detailed studies of thermodynamic properties by conventional methods.

The use of recording differential techniques has greatly emphasized two items of importance. 1) The solid phase polymorphism of many mesophase forming materials is as complex as the mesophase behavior. Indeed, this statement may be made in general about organic compounds: solid phase polymorphism is very common. 2) Absolute purity and types of impurity play very important roles in the type, transition temperature and range of mesophase formation. To some extent, this was realized by most early careful workers, but these methods did not drive the point home as effectively or dramatically as differential measurements.

The purpose of this paper is to outline the thermodynamic data which are presently available for homologous series of liquid crystal forming materials. Materials will be listed only when calorimetric data are available. It is hoped that this outline will furnish a framework for the further study and systemization of the increasing body of information appearing monthly. It will become obvious to the reader that certain general statements relate the thermodynamic constants of all liquid crystal forming materials. Without a doubt these general statements will be revised and grow more powerful as additional information from other sources is added in the future. A more complete listing of thermal properties will shortly become available as volume 2 of reference 3.

TECHNIQUES OF MEASUREMENT

Arnold and coworkers have made a large number of measurements using classical adiabatic calorimetry.[5,6] However, the bulk of the thermodynamic data presently available has been obtained by dynamic calorimetry. By the very nature of the instrumentation dynamic methods are much less accurate than adiabatic calorimetry. The expected accuracy is usually not much better than ±1% and in some cases only ±10%. None the less, the information is useful for general comparisons, so long as the limitations are recognized. Kreutzer and Kast[7] employed dual ice calorimetry which is intermediate in precision between adiabatic and scanning calorimetry.

The techniques of DTA have been described in detail elsewhere.[8] Briefly, a sample and reference material are heated at some linear rate. The absolute temperature and the differential temperature between sample and reference are recorded. The area beneath the differential curve is related to calories via calibration with a material of known heat of fusion. Since the area is due to a temperature difference, factors such as sample and instrument heat capacity are important. Temperatures, heats of transition and heat capacity may be determined from the curves to about ±1% given adequate calibration.

DSC also involves the comparison of the sample with an inert reference during a dynamic heating or cooling program. However, instead of a temperature difference being permitted, heat is added via the filaments, to keep the sample and reference in balance. To a first approximation, this method is a kind of scanning adiabatic calorimetry.

Obviously, both scanning methods are measuring the same phenomena. However, it is a serious logical error to confuse the two techniques. When adequately calibrated and carefully employed the two methods give comparable results. Indeed, under favorable circumstances DSC data compare closely to data from the highly precise calorimetry of Arnold, see Fig. 2, and calculated values. Serious errors may be expected for transitions which do not reach equilibrium quickly. Fortunately, most liquid crystal transitions are rapid.

Other methods for the evaluation of transition heat have been employed occasionally for liquid crystals. These methods will be mentioned in passing when particular compounds are discussed.

THERMODYNAMIC DATA

The data in this collection are gathered from various articles in the literature. The original workers reported the results in a variety of units, i.e., cal/g, cal/mole, joules/mole and degrees Kelvin and centigrade. For the sake of unity of presentation all of the diverse units have been converted to calories/mole for heat of transition and calories/mole/°K for entropy of transition. In some cases various authors have given several values for transition heat depending upon the point taken as the start of transition. The effect of purity is very large on the heat and temperature of transition.[10,11]

When comparing transition heats measured by various methods or by the same method by different authors it is not uncommon to find variations as large as 10%. This does not imply less precision on the part of one worker over another or necessarily imply differences in sample purity. The usual source of the "error" is how much of the pre-transition effect is included in the measured transition heat. Liquid crystal transitions are certainly not thermodynamically simple, McCullough[12] has identified at least six characteristic non-isothermal melting and solid-solid transitions on the basis of heat capacity-temperature curves of organic compounds. From the available heat capacity curves of liquid crystal forming materials it is obvious that several of the above non-isothermal processes are represented in nematic and cholesteric systems.[13,14]

However, it is one thing to recognize a problem and yet another to arrive at a satisfactory solution. For a non-isothermal process the "pretransition energy" must be included and properly belongs with the main transition energy—especially if entropy of transition is to be used to judge order changes in a system. Given two sets of measurements on samples of equivalent purity, the set involving the largest energy change is more likely to be correct. The statement about equivalent purity is the axis of the argument, for impurities acting as solvents can bring about non-isothermal melting in systems which should be isothermal. It is very useful to know not only the

177

temperature of maximum transition rate (T_m) but also the temperature range considered in evaluating the heat of transition. For the present, this sort of information is unavailable in many cases. Hopefully, transition range will be included in more reports in the future. As for the present study, transition entropies will be reported directly as given in the literature with a minimum of speculation concerning possible pretransition effects.

A sampling of the various compound structures for which thermodynamic information is available is given in Table I. In most cases at least ten members of a homologous series have been made and studied. In series which exhibit properties of interest to liquid crystal displays or thermal indication as many as twenty-five members of the series have been studied. This list should be regarded as an underline only as the totals and types continue to change. The letters, A, B, C, etc., will be used where convenient in this text for reference.

A quick examination of Table I reveals that liquid crystal forming materials are usually rod-shaped and rigid (A to F) or have some unusual symmetry, G. The rigid rod shaped materials may exhibit smectic and nematic mesophases in any combination with the nematic form being the highest melting. A large number of smectic forms have been defined, see Fig. 3. Compounds belonging to the cholesteryl group, G, exhibit smectic and cholesteric mesophases. The nematic mesophase is excluded in pure materials since the cholesteric is a special case. To the present authors knowledge only a single smectic and single cholesteric transition is observable in G-type materials. That is, the transition sequence is never larger than solid, smectic, cholesteric, isotropic liquid. Two or three smectics are not observed as may be the case with any of the other materials.

Monotropism

Mesophases may also be monotropic. That is, form only on cooling the isotropic liquid. This is usually due to the extent of the crystalline order being great enough to preserve the true solid above the range of mesophase thermal stability.[16,17] When impurities are added to many monotropic systems, the melting of the solid is depressed sufficiently to make the mesophases non-monotropic. Cholesteryl palmitate is a good case in point. In ultra-pure material the transition path is as follows.[18]

However, in the presence of a trace of antioxident the melting of the crystal is depressed giving:[19]

Crystal ⇌ Smectic ⇌ Cholesteric ⇌ Isotropic Liquid

Any or all the mesophases may be monotropic. Solid phase polymorphism may also determine monotropism, and the phase from which crystallization occurs

178

determine the nature of the solid phase formed. Cholesteryl formate and propionate are instructive of this phenomenon.

Using a formate ester sample with a DSC purity of 99.86% (98.67% using the method of Davis et al.[20]) the following transitions were noted using a polarizing microscope and hot stage:[21]

1. On heating, the solid conversion to the isotropic liquid was noted at 97.0°C.

2. The stage temperature was lowered to 95.7°C and a prismatic solid formed.

3. Rapid cooling of the melt on the slide produced a cholesteric mesophase.

4. By trial and error, a stage temperature, 57.2°C, was found at which the mesophase would neither melt to the isotropic liquid nor the above solid form rapidly.

5. After several changes of the "moss-like" cholesteric texture, the mesophase became isotropic at 59.6°C.

6. If the mesophase is cooled to below 50°C a new, spherulitic mesophase forms.

The above observations suggest the following path of phase transition:

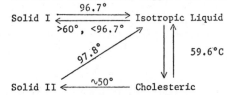

Note, solid I melting followed by slow cooling indicates no mesophase. The enthalpy difference between solid I and solid II is at present not known. The present authors suggest that it is less than 1 kcal/mole.

Three sets of calorimetric data are available. Agreement between three independent studies of the solid → isotropic liquid transition is excellent, 2% in terms of ΔS. The difference between the two measurements of the very small mesophase transition is ∿9% in terms of ΔS.

The propionate ester is the first of the aliphatic ester series that all workers agree forms a mesophase--and that on heating as well as cooling. However, precise scanning calorimetry has demonstrated the existence of two solid phases with different melting points.[22] Microscopy has verified two solid forms--solid I, spherulitic and solid II, needle-like. The following recrystallization path was suggested:[22]

```
Solid I ⇌ Cholesteric ⇌ Isotropic Liquid
                    ↑
           ∿88°  ∥  98.0°
                    ↓
             Solid II
```

No direct conversion from solid I to solid II without the mesophase was
observed. Five sets of thermodynamic data and two sets of transition
temperatures are known.[23-27] The variation in transition temperature is
probably due to differing amounts of solids I and II in individual samples.

Materials Exhibiting the Cholesteric Mesophase

All compounds with structure G exhibit a cholesteric mesophase (hence,
the name). This characteristic twisted or glide-plane structure has also
been found in d or ℓ nematic mesophase forms containing an asymmetric carbon
atom in the side chain. The racemic d-ℓ mixtures are normal nematics. In
particular the compound

which belongs to type E or Schiff's bases has been studied in both the d and
d-ℓ mixture form.[28,29] The d form shows a good cholesteric at 99.5°C
(monotropic) with an entropy of 0.27 cal/mole/°K.[29] The solid to isotropic
liquid transition occurs at 99.6°C with a ΔS of 17.5 cal/mole/°K. The d-ℓ
form is not monotropic with the solid to nematic transition at 100°C and ΔS
= 17.3 cal/mole/°K. The nematic to isotropic liquid transition occurs at
109°C with a ΔS of 0.24 cal/mole/°K. The differences in the mesophases are
probably real.

Cholesterol Mesogens

More conventional, type G, cholesteric mesogens have been studied in
great detail. For example the n-alkyl acid esters have been evaluated up to
twenty one carbon numbers, and for most of the series by at least four
workers.[23-27] The transition entropy for the solid to mesophase (or isotropic
liquid in monotropic cases) process is shown as a function of acid chain
carbon number in Fig. 4. The scatter below C_5 is probably due to polymorphism
in the solid phase. The sharp discontinuity between C_8 and C_9 has been
reported by most workers, and from x-ray diffraction studies appears to be
due to a fundamental change in molecular packing.[30] Above C_{10} there is some
evidence for even-odd alternation in entropy.

The same kind of plot for the mesophase transitions is given in Fig. 5.
From C_1 to C_8 the cholesteric → isotropic liquid entropy increases in a
relatively smooth manner. A sharp break in this trend occurs at C_9 when a
lower but more linear trend is established up to C_{20}. Also, at C_9 the smectic

180

mesophase appears. The entropy of the smectic to cholesteric transition increases even more rapidly with additional carbon numbers. This would be expected from the closer association of molecules in the smectic mesophase.

A detailed study using non-alkane acid esters has shown that the transition entropies of cholesteryl esters is sensitive only to the length of the ester group.[31] When the length of the ester in Angstroms is plotted with entropy, such diverse materials as the benzoate, adipate, and tetrafluorobutyrate esters fall on the same smooth relationship.[31] Indeed the ester functionality is not necessary to preserve. Marker's Acid (the positional analogue of cholesteryl formate) has almost identical solid and mesophase (cholesteric) transition entropies although hydrogen bonding in the acid translates the transition temperatures upward by more than 150°C.[32]

| Formate Ester | Marker's Acid |

Indeed, for the cholesterol mesogens, geometry appears to be the only important term in defining the entropy change involved in any transition. Ennulat et al. have investigated the thio esters, carbonates and thio carbonates of cholesterol.[23,33-35] Allowing for some small differences in chain length brought about by the thio-substitution, the analogues are almost identical to their oxygen counterparts. The thio-esters do exhibit the smectic mesophase at C_8.

Other Mesogens

These materials obviously form the bulk of the compounds investigated. They are characterized by a rod-like shape and more or less strong dipolar interactions. Although in Table I the central bridge between rings has been stressed, the end group substitution has profound influence on the temperature and heat of the various transitions.

Two extreme cases have been chosen as examples. The azoxy benzenes and azoxy cinnamates belong to group A of Table I, but the cinnamate has much more polar end groups. The thermodynamic picture is correspondingly vastly different. Arnold has carried out precision calorimetry on the methyl through dodecyl values of the side chain[6] for the symmetrical dialkoxyazoxybenzenes as well as octadecyl. His values for transition heats are shown in Fig. 6. In addition, the first two members of the series, p-azoxyanisole, see Table II, p-azoxyphenetole and p,p'-dihexyloxyazoxybenzene have been studied by a number of other workers.

Arnold's calorimetric data present a remarkably scattered picture for the transition from the solid to the mesophase, Fig. 6. No regular order is apparent. This is somewhat surprising in consideration of the regular order which is evident for the nematic → isotropic, smectic → nematic and smectic → isotropic transitions in the same study. A possible explanation for such erratic entropy of transition values is offered in Table II. Chow and Martire

confirmed the existance of two separate and distinct solid phases of
p-azoxyanisole ∼99.7 mole%, by scanning calorimetry.[40] Although solid II,
the high temperature form, is metastable at room temperature; it can exist
in various mixtures with solid I. The formation of two or more solid phases
is very common in mesophase forming materials. In considering the entropy
of the transition solid → mesophase it is necessary to use the total entropy,
i.e., the summed entropy of transition for all crystal forms which are stable
between 0°K and the transition temperature. Bondi has adequately demonstrated
this for other organic transitions.[42] Therefore, the present authors suggest
that most of the materials investigated by Arnold were complicated by solid
phase polymorphism. The relationship between transition entropy and alkyl
chain length should be regular if all crystal forms were accounted for. The
collection of data in Table II for the solid to nematic mesophase transition
of p-azoxyanisole is scattered for probably the same reason. It is
interesting to note that the values for nematic → isotropic liquid transition
as measured by eight workers are in very close agreement. By elimination of
the extremes the value for this transition is between 0.403 and 0.453
cal/mole/°K. Thus, errors due to the small entropy of transition
notwithstanding, the nematic isotropic transition is better defined than
transition involving the solid phase, i.e., polymorphism does not complicate
the nematic mesophase. In the calorimetric study, Fig. 6, the nematic
isotropic transition shows even-odd alternation up to the C_8 side chain.

Chow and Martire measured the thermodynamic properties of the C_6 side
chain material, 99.6 mole%.[40] The entropy of the solid → nematic transition
was found to be 29.4 cal/mole/°K (heat of transition 10.4 kcal/mole). This
compares to Arnold's value of 27.92 cal/mole/°K. The nematic → isotropic
liquid transition was found by the same authors to have an entropy of 0.895
cal/mole/°K (heat of transition 0.359 kcal/mole). This compares closely with
Arnold's extrapolated value of 0.898 cal/mole/°K. In addition, Chow and
Martire noted the presence of a smectic mesophase very close to the solid
transition.[40] They were unable to determine the heat of transition due to
poor resolution on the DSC trace.

In general the results of DTA and DSC on the mesophase → isotropic liquid
transition compare more closely with Arnold's "extrapolated" values of
transition heat than with the quantities derived by more straightforward
measurements. The extrapolated values included pretransitional heat capacity
changes. The method of baseline construction in DTA and DSC accounts for
the pretransition inclusion. Martin and Muller have discussed this
phenomenon.[38]

In the homologous series of p,p'-dialkoxyazoxybenzenes the alkyl side
chains do not appear to contribute significantly to the order of the nematic
system up to C_6. Below C_6 the alkyl side chains are probably not laterally
associated and may determine which of two possible lateral arrangements is
adopted by neighboring azoxybenzene nuclei in order to account for the
odd-even effect. At C_6 the configurational geometry of the side chains
becomes such that they are closely associated and contribute to the order of
the system. Undoubtedly the appearance at either C_6 or C_7 of a smectic
mesophase is due to interalkyl chain order. Beyond C_{10} the interchain order
is so elaborated that the stability range of the nematic mesophase is exceeded

by the time the system has sufficient energy to separate the alkyl chains. Thus, the nematic mesophase vanishes from the phase diagram.

Between C_{12} and C_{18} a new type of order appears, and at C_{18} two smectic mesophases appear, the smectic B and smectic C of Saupe.[43] The C_{18} compound is one of the few materials giving two smectic mesophases for which calorimetric information is available. It may be characteristic of the smectic B → smectic C transition that it is smaller in entropy change than the smectic C → isotropic liquid transition. Logically, this is difficult to visualize. If the same chain segments are involved in both transitions, then the transition entropies should be approximately equal--not a factor of two different, 6.79 cal/mole/°K versus 13.9 cal/mole/°K. It is possible to speculate that only half the alkyl chains are involved in the smectic B → smectic C transition. This sort of speculation is predicted on the fact that the smectic B modification is not a metastable form.

Arnold has presented a heat capacity versus temperature curve for the dioctadecyl material.[44] The smectic B form shows little pretransition heat capacity anomaly where as the smectic C modification shows a relatively large pretransition effect (as does the solid). The heat capacity in the mesophase range is unusually high; well above that of both the isotropic liquid and the solid phases. This is analogous to the heat capacity phenomena noted for the smectic and cholesteric mesophase of cholesteryl myristate.[13]

Arnold, Demus, Kock, Nelles, and Sachmann[45] have prepared and thermally characterized a series of compounds with the general structure:

$$C_nH_{2n+1}\text{-O-}\overset{\overset{O}{\parallel}}{C}\text{-}\underset{\underset{CH_3}{|}}{C}=CH-\langle O \rangle -N=\overset{+}{\underset{\underset{O}{|}}{N}}-\langle O \rangle -CH=\underset{\underset{CH_3}{|}}{C}\text{-}\overset{\overset{O}{\parallel}}{C}\text{-O-}C_nH_{2n+1}$$

with n = 2,3,4,5,6,7,8,9 and 11. These compounds represent a symmetrical form of the Schiff's base p-aminocinnamic acid esters described by Leclercq et al.[28] with a much increased dipole due to the nitroso bridge. This should be comparable to the previously discussed 4,4'-di-n-alkoxyazoxybenzenes.[6] The entropic data are shown in Figs. 7 and 8.

The ethyl and propyl diesters have two mesophases: smectic A and nematic. The nematic mesophase is not exhibited by the next four members of the series. Only the smectic A mesophase appears. From the octyl to dodecyl diester two mesophases appear: smectic A and smectic C. Complete heat capacity curves of the mesophase range of the didecyl and didodecyl esters have been presented.[46]

Figure 7 shows the solid to mesophase transition entropy as a function of ester carbon number. Although the scatter is bad, there is a general upward trend with increasing ester chain length. As with the p,p'-n-dialkoxy azoxybenzenes this scatter is probably due to solid phase polymorphy.

Figure 8 indicates that the mesophase entropies are somewhat more regular. Contrary to other smectic series which have been studied, there is little

183

evidence of even-odd carbon number alternation. The series increases along a smooth curve from diethyl to didodecyl. The entropies of nematic, smectic A and smectic B are shown individually and totaled in Fig. 8. The totaled form would be expected to be more significant when all mesophases are not monotropic. The entropy increase for this series is less rapid with increasing chain length than with the dialkoxy azoxybenzenes. However, it is more regular. The additional chain length in the cinnamic acid link probably does not contribute to the smectic order. If a contribution were made the entropy of the didodecyl compounds of azoxy-α-methylcinnamic acid and azoxybenzenes should be comparable with the cinnamate the larger. This is not the case.

CONCLUSIONS

From this brief summary of known mesophase thermodynamics it is apparent that much can be learned from this complex picture. In theory, transition entropy and heat capacity define molecular order and can be used to describe in both words and mathematics exactly how molecules behave in such diverse states as solid and liquid. X-ray studies have given us the outline of matter just as does a city map give us the frame of a city. The frame is just that, not filled in with the patterns of flow and movement for which the city really exists. The same is true of a crystallographic approach to matter; movement of the units is neglected because of the static nature of the method. Thermodynamics gives a picture (albeit complex) of the movement of molecules and the effects of size, shape and dipole interaction. Liquid crystal forming materials have probably the most completely explored thermal picture (with the possible exception of the n-alkanes) of any group of chemicals. The body of data available on homologous series is very large and should be helpful in disentangling dipolar from other effects. The cholesteric series is almost a pure function of geometry as contrasted to nematic mesogens. Liquid crystals constitute an important interface, and it has been at interfaces at which the largest advances of our knowledge of the universe have been made.

REFERENCES

1. G. H. Brown and W. G. Shaw, Chem. Rev. 57, 1049 (1957).

2. J. L. Fergason, Sci. Am. 211, 76 (1964).

3. G. W. Gray, Molecular Structure and Properties of Liquid Crystals, Academic Press, Inc., New York (1962), pp. 131-133.

4. J. L. Fergason, T. R. Taylor, and T. B. Harsh, Electro Technology 85, 41 (1970).

5. H. Arnold, Z. Physik. Chem. (Leipzig) 225. 45 (1964).

6. H. Arnold, Z. Physik. Chem. (Leipzig) 226, 146 (1964).

7. K. Kreutzer and W. Kast, Naturwissenschaften 25, 233 (1937).

8. E. M. Barrall and J. F. Johnson, Techniques and Methods of Polymer Evaluation, Vol. I, P. E. Slade and L. T. Jenkins, Eds., M. Dekker, N.Y. (1966), pp. 1-40.

9. A. Torgalkar, R. S. Porter, E. M. Barrall, and J. F. Johnson, J. Chem. Phys. 48, 3897 (1968).

10. R. D. Ennulat, Analytical Calorimetry, Vol. I, R. S. Porter and J. F. Johnson, Eds., Plenum Press, N.Y. (1968), p. 219.

11. E. M. Barrall and M. H. Vogel, Thermochem. Acta 1, 127 (1970).

12. J. P. McCullough, Pure, Appl. Chem. 2, 221 (1961).

13. E. M. Barrall, R. S. Porter, and J. F. Johnson, J. Phys. Chem. 71, 895 (1967).

14. H. Arnold and P. Roediger, Z. Physik. Chem. (Leipzig) 239, 283 (1968).

15. G. H. Brown, J. W. Doane, and V. D. Neff, Critical Reviews in Solid State Sciences, 303, September (1970).

16. W. R. Young, I. Haller, and A. Aviram, Program Abstract, Third International Liquid Crystal Conference, Aug. 24-28, 1970, Berlin, p. 136.

17. J. S. Dave and M. J. S. Dewar, J. Chem. Soc., 4305 (1955).

18. E. M. Barrall, R. S. Porter and J. F. Johnson, J. Phys. Chem. 71, 1224 (1967).

19. M. Vogel, E. M. Barrall, and C. P. Mignosa, Molecular and Liquid Cryst. 15, 49 (1971).

20. G. J. Davis, R. S. Porter, and E. M. Barrall, Molecular and Liquid Cryst. 10, 1 (1970).

21. E. M. Barrall and R. S. Porter, work in progress.

22. M. J. Vogel, E. M. Barrall, and C. P. Mignosa, Second Symposium on Ordered Fluids and Liquid Crystals, J. F. Johnson and R. S. Porter, Eds., Plenum Press, N.Y. (1970), p. 333.

23. R. D. Ennulat, Molecular and Liquid Cryst. 8, 247 (1969).

24. E. M. Barrall, R. S. Porter, and J. F. Johnson, J. Phys. Chem. 71, 1224 (1967).

25. G. J. Davis and R. S. Porter, in press, J. Thermal Analysis.

26. P. J. Sell and A. W. Newman, Z. Physik. Chem. (N.F.) 65, 13 (1969).

27. G. W. Gray, J. Chem. Soc. (London), 3733 (1956).

28. M. Leclercq, J. Billard, and J. Jacques, Molecular and Liquid Cryst. 10, 429 (1970).

29. E. M. Barrall, R. S. Porter, and J. F. Johnson, Molecular Cryst. 3, 299 (1968).

30. J. H. Wendorff and F. P. Price, J. Phys. Chem., in press.

31. E. M. Barrall, J. F. Johnson, and R. S. Porter, Molecular and Liquid Cryst. 8, 27 (1969).

32. W. R. Young, E. M. Barrall, and A. Aviram, Analytical Calorimetry, Vol. 2, J. F. Johnson and R. S. Porter, Eds., Plenum Press, N.Y. (1970), p. 113.

33. W. Elser, Molecular and Liquid Cryst. 8, 219 (1969).

34. W. Elser, Molecular and Liquid Cryst. 2, 1 (1967).

35. W. Elser and R. D. Ennulat, J. Phys. Chem. 74, 1545 (1970).

36. E. M. Barrall, R. S. Porter, and J. F. Johnson, J. Phys. Chem. 68, 2801 (1964).

37. R. Schenk, Kristalline Flüssigkeiten und Flüssigekristalle," W. Engelmann Verlag, Leipzig (1905), pp. 84-89.

38. H. Martin and F. H. Müller, Kolloid-Z.u.Z. für Polymere 187, No. 2, 107 (1963).

39. K. Kreutzer, Ann. Physik 33, 192 (1938).

40. L. C. Chow and D. E. Martire, J. Phys. Chem. <u>73</u>, 1127 (1969).

41. M. Leclercq, J. Billard, and J. Jacques, Compt. Rend. <u>264</u>, 1789 (1967).

42. A. Bondi, Chem. Rev. <u>67</u>, 565 (1967).

43. A. Saupe, Molecular Cryst. <u>7</u>, 59 (1969).

44. H. Arnold, E. G. El-Jazairi, and H. König, Z. Physik. Chem. <u>234</u>, 401 (1967).

45. H. Arnold, D. Demus, H.-J. Koch, A. Nelles, and H. Sackmann, Z. Physik. Chem. (Leipzig) <u>240</u>, 185 (1969).

46. H. Arnold, J. Jacobs, and O. Sonntag, Z. Physik. Chem. (Leipzig) <u>240</u>, 177 (1969).

Table I

Some Structures Which Yield Liquid Crystal Families of Known
Thermodynamic Properties

A

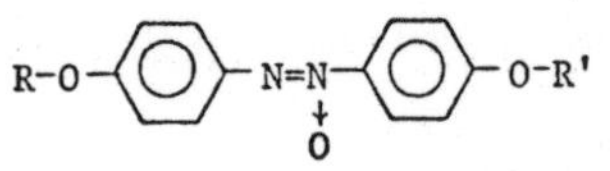

4,4'-di-n-alkoxyazoxybenzenes

B

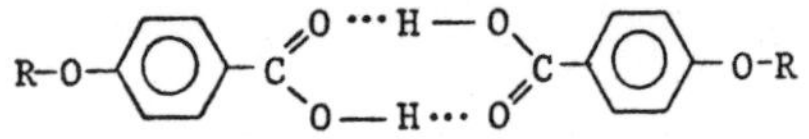

p-alkoxybenzoic acids

C

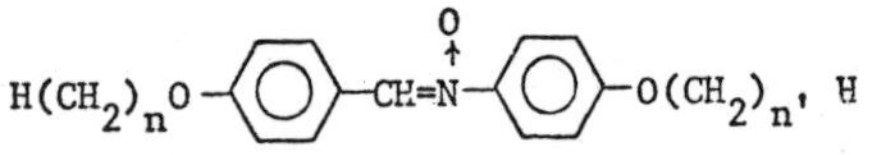

4,4'-di-n-alkoxyphenylnitrones

D

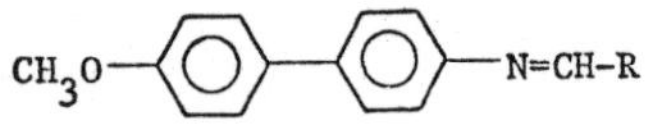

4-benzylidene-amino-4'-methoxybiphenyls
(R = aromatic or heterocyclic)

E

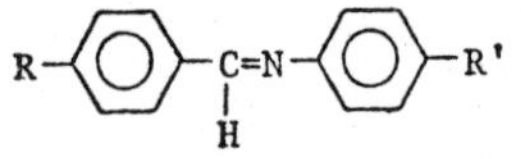

Schiff's bases

Table I - Continued

F   R—⬡—C≡C—⬡—R

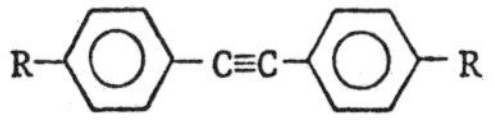

diphenylacetylides

G

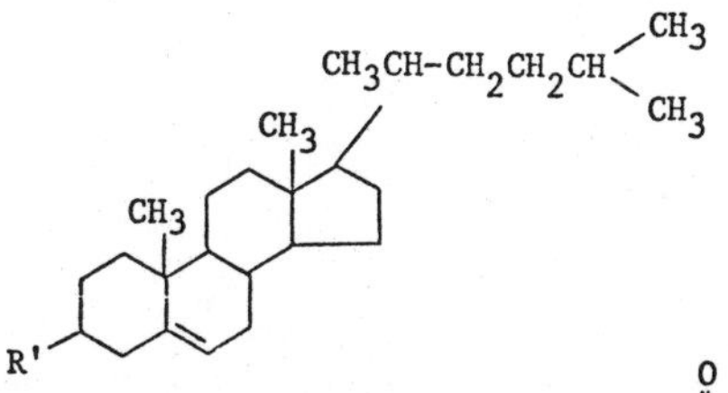

$$CH_3CH-CH_2CH_2CH\begin{matrix} CH_3 \\ \\ CH_3 \end{matrix}$$

cholesteryl esters

R'

carbonates

$$\overset{O}{\underset{\|}{R-C-O-}}$$

$$\overset{O}{\underset{\|}{R-O-C-O-}}$$

S-alkyl carbonates $$\overset{O}{\underset{\|}{R-S-C-O-}}$$

ω-phenyl alkanoates $$\phi-(CH_2)_n\overset{O}{\underset{\|}{-C-O-}}$$

Thiocholesteryl-ω-phenylalkanoates $$\phi(CH_2)_n\overset{O}{\underset{\|}{C-S-}}$$

halides $C\ell-, Br-, I-$

189

Table II

A Comparison of the Thermodynamic Data Available on the
Transitions of p,p'-azoxyanisole

Refer.	Transition Type	Trans Temp °C	Trans Temp °K	Trans Heat Cal/Mole	Trans Entropy Cal/Mole/°K
36	Solid → Nematic	117.6	390.6	7440	19.0
	Nematic → Isotropic	133.9	406.9	176	0.432
37	Solid → Nematic	118	391	7700	19.7
	Nematic → Isotropic	132	405	176	0.434
38	Solid → Nematic	117	390	7280	18.7
	Nematic → Isotropic	128 onset	401	178	0.444
	Nematic → Isotropic	132 sharp break	405	178	0.440
37	Nematic → Isotropic	132	405	176	0.434
37	Nematic → Isotropic	132	405	183	0.453
7,39	Nematic → Isotropic	132	404	462	1.14
40	Solid → Nematic	117.5	390.5	7260	18.6
	Nematic → Isotropic	134.2	407.2	181	0.444
	Solid II → Nematic	104.4	377.4	5630	14.9
	Nematic → Solid II	76	349	7230	20.7
	Solid II → Solid I	61	334	1030	3.09
6	Solid → Nematic	118.2	391.2	7067	18.07
	Nematic → Isotropic	135.3	408.3	137.2	0.3360
	Nematic → Isotropic	135.3 extrapolated	408.3	164.9	0.4039
41	Solid → Nematic	117.5	390.5	6800	17.4
	Nematic → Isotropic	134.0	407.0	150	0.37

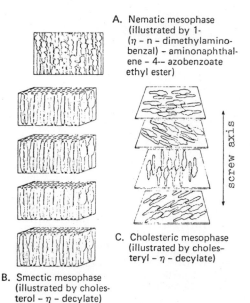

A. Nematic mesophase
(illustrated by 1-
(η – n – dimethylamino-
benzal) – aminonaphthal-
ene – 4-- azobenzoate
ethyl ester)

C. Cholesteric mesophase
(illustrated by choles-
teryl – η – decylate)

B. Smectic mesophase
(illustrated by choles-
terol – η – decylate)

Fig. 1 Three types of thermodynamic mesophases.

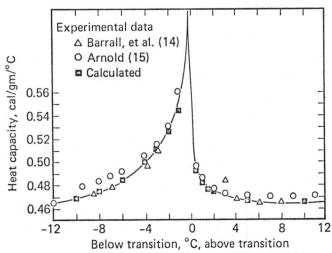

Fig. 2 A comparison of experimental (calorimetry and DSC) and calculated heat capacity data[9].

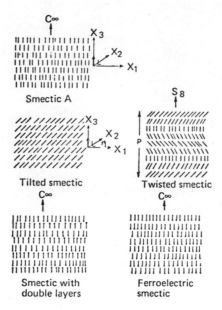

Fig. 3 Schematic representation of several smectic liquid crystal geometries.

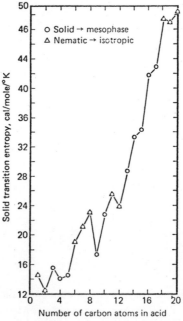

Fig. 4 Effects of acid carbon number on solid → mesophase O or solid → isotropic liquid X transition entropy of cholesteryl esters.

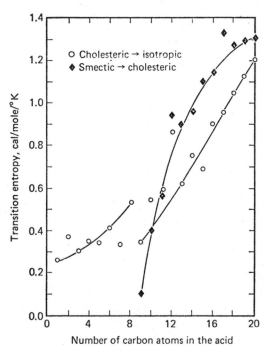

Fig. 5   Effect of acid carbon number on mesophase transition
         entropy of cholesteryl esters.

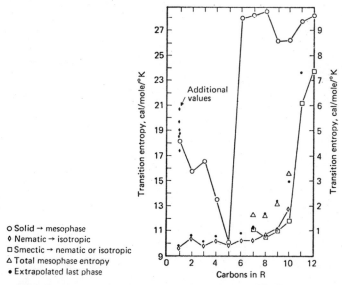

Fig. 6   Transition entropies for some p,p-n-dialkoxy-
         azoxybenzenes.

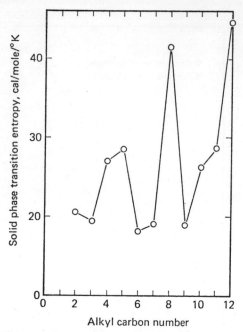

Fig. 7 Effect of alkyl chain length on the solid to
mesophase transition entropy for a series of
4-4'-azoxy-α-methyl cinnamic acid di-n-alkyl
esters.

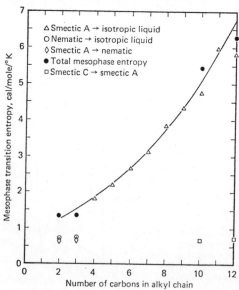

Fig. 8 Effect of the alkyl chain length on mesophase
transition entropies for a series of 4,4'-
azoxy-α-methyl cinnamic acid di-n-alkyl
esters.

STUDIES ON CATALYSTS AND CATALYTIC REACTIONS
BY GAS-FLOW DTA AND HIGH-PRESSURE DTA

T. Ishii

Division of Applied Chemistry, Faculty of Engineering
Hokkaido University, Sapporo, Japan

ABSTRACT

Gas-flow DTA (applicable from static atmosphere to fluidized bed state) and high-pressure DTA (applicable up to 200 kg/cm^2 and 500°C) were developed and applied to the following studies. The features of the apparatus and the catalytic action under working conditions are described.
(I) Gas-flow DTA : 1) thermal dehydration of $CuSO_4 \cdot 5H_2O$, oxidation of UO_2 and decomposition of $MgCl_2 \cdot 6H_2O$, 2) DTA studies of V_2O_5-M_2SO_4 catalysts (M = Li, Na, K and Cs) for SO_2 oxidation, 3) thermal decomposition of $KClO_4$ by Fe_2O_3 catalysts with different preparing histories.
(II) High-pressure DTA : 1) liquid-phase high-pressure hydrogenation of aromatic substances, benzene, phenol and diphenylether, by commercial catalysts, 2) high pressure coal hydrogenation by various catalysts, Fe_2O_3, Fe_2O_3-S, red mud, red mud-S, FeS, $ZnCl_2$, ZnO, ZnS, $SnCl_2 \cdot 2H_2O$, SnO_2, SnS, $(NH_4)_6Mo_7O_{27}$.

INTRODUCTION

In carring out the investigation of a chemical reaction the experimental results are generally obtained in the form of the composition changes of reaction mixture as a function of time, temperature, and/or atmosphere, etc. In this case, it is common knowledge that these data have to be procured by means of investigations with the suitable experimental reactor which is operated isothermally.

Differential thermal analysis (DTA) technique differs essentially from such conventional isothermal methods, and is a nonisothermal technique which utilizes the energy evolved or absorbed by characteristic reaction when the test sample is placed in an environment of rising temperature at constant rate. A striking advantage of DTA tests is that the procedure requires only a small fraction of the time and the reaction processes over the wide temperature range can be directly followed by the observation of the DTA curve. In applying this technique to the studies of catalysts and catalytic reactions, it is possible to observe directly the thermal behavior of catalysts themselves and the reacting system under working state of the catalysts.

However, it is inevitable that the quantitative analysis of DTA curve is very complex because it is a dynamic investigating

method. Furthermore, other techniques such as X-ray diffraction,
IR analysis, chemical analysis, gas analysis, etc. must be used
in conjunction with DTA studies to understand completely the
physical or chemical meaning of the thermal deflection on DTA
curve.
On the standpoint mentioned above, DTA techniques have been
applied extensively to the various type of chemical reactions in
our laboratory. In this case, the controlled-atmosphere or the
controlled-pressure DTA apparatus needed to be used. So, simple
gas-flow DTA apparatus with a flowing gas technique in which gas
flows through the reference and the sample powder beds, and high-
pressure DTA apparatus operated under high pressure had been
developed, and have been continued to study the reactions between
gas, liquid and solid phases.

APPLICATION OF GAS-FLOW DTA APPARATUS

Gas-flow DTA apparatus (1)

Fig.1-A and B show the schematic diagrams of gas-flow DTA
and of DTA cell, respectively. DTA apparatus consisted of two
quartz tubes (id.=10mm) placed vertically in the furnace, one
was used as reference cell and the other as sample cell, in which
gas flows through the reference and sample powders during test.
Pulverized sample and reference material were kept at fixed position in the tube by underlying loosely packed quartz wool.
Chromel-Alumel thermocouples (d=0.3mm) were inserted in each
of these tubes to enable the measurement of differential temperature. The thermocouples were protected by fine quartz tube
(od.=4.5mm) from corrosive atmospheric gases such as Cl_2, SO_2,
SO_3, etc. The heating rate of the furnace was controlled by a
ECP-51B type program controller (Ohkura Electronic Ins.) and the
differential temperature was recorded on a 25SB4-IR-6 milivolt
recorder (Ohkura) through AM-1001B type micro volt meter (Ohkura).
One gram of Al_2O_3 (-150 mesh, Merck) calcined at 1290°C for 3 hrs
was used as a reference material. The temperature ranges up to
about 1000°C were used for application of this apparatus, and
strong corrosive gases such as Cl_2, SO_2, SO_3, HCl, etc. could
flow at a wide range of gas flow rates, 0 ml/min (static technique) to 300 ml/min (fluidized-bed technique).
In order to analyse the changes of sample in the course of
DTA studies, the heating sample was quenched rapidly to room
temperature by the method of throwing water on the sample cell
just after it was taken out of the furnace.
It must be kept in mind for a successful operation of this
apparatus that heating rate and gas flow rate must be maintained
constant during tests.

Influences of atmosphere and gas flow rate on DTA curve

Little attention has been given to the influences of the
composition and flowing condition of gas phase surrounding the
sample on the resultant DTA curves. In general, the DTA curve
obtained by using the static gas technique differs from the curve
for the same powder in which the gas phase is passing through
the sample bed. In such a case, the observed differences between
the two may give a very good clue in the consideration of reaction

mechanisms.

(1) Thermal dehydration of $CuSO_4 \cdot 5H_2O$ (1)
Fig.2 shows the DTA curves for thermal dehydration of $CuSO_4 \cdot 5H_2O$ at various air flow rates changing from static state to fluidized bed condition at gas flow rate of 250 ml/min. Three endothermic peaks, prak 1, 2 and 3, correspond to dehydration of $2H_2O$, $2H_2O$ and H_2O, respectively. All the peaks were shifted significantly to the lower temperatures with increasing gas flow rate in a similar manner. This result shows that the over-all dehydration rate was promoted by decreasing the resistance of mass transfer at the external particle surface in the flowing gas condition. On the other hand, the effect of air flow rate on peak area for dehydration of $CuSO_4 \cdot 5H_2O$ can be regarded as insignificant as shown in Fig.3.

(2) Oxidation of UO_2 (1)
Fig.4 shows the DTA curves for oxidation of UO_2 in air. Exothermic peak 1 and 2 correspond to the reaction for $UO_2 \rightarrow U_3O_7$ and $U_3O_7 \rightarrow U_3O_8$, respectively. The shape of peak 2 was strongly affected by air flow rate, while the shape of peak 1 and the temperatures of peak 1 and 2 were only slightly affected.
No simple explanation for the thermal behaviors observed can be offered at the present time, but it may probably be suggested, at least in part, that the rate determing step is not the diffusion process of oxygen to particle surface from bulk air stream, and the reaction mechanism of peak 1 and 2 could be different.
The activation energies calculated by Kissinger's method, 23.1 kcal/mol for the peak 1 and 38.7 kcal/mol for the peak 2, agreed with the values reported in the previous papers. No significant difference of activation energy between static and fluidized conditions was observed for this reactions.

(3) Decomposition of $MgCl_2 \cdot 6H_2O$ (2)
The process of thermal decomposition of $MgCl_2 \cdot 6H_2O$ is complicated and greatly influenced by the surrounding atmosphere.
Fig.5 shows the DTA curves of $MgCl_2 \cdot 6H_2O$ in flowing air of various flow rates. All peak temperatures shifted to lower temperature with increasing flow rate in the similar manner.
Fig.6 shows the results in N_2-steam atmospheres. Endothermic peaks of dehydration, peak 2-5, and peak 6 were shifted to higher temperature site compared with in dry N_2 atmosphere. Shifting of peak 2-5 is due to the action of steam atmosphere, and that of peak 6 may be due to HCl liberated at peak 5.
Activation energies estimated by Kissinger's method under fluidized bed condition in air were 12.6 lcal/mol for $MgCl_2 \cdot 4H_2O$ $\rightarrow MgCl_2 \cdot 2H_2O + 2H_2O$ (peak 3) and 57.8 kcal/mol for $MgOHCl \rightarrow MgO + HCl$ (peak 6).

DTA studies of vanadium catalysts for SO_2 oxidation (3)

Industrial vanadium catalysts generally consist of V_2O_5, alkali salts promoter and diatomaceous earth carrier. In a working state, these catalysts are in the molten state and have a complex composition resulting from the interaction between catalyst and gaseous reactants.
The thermal behaviors of vanadium catalyst with alkali

salts were studied in the atmospheric condition similar to the industrial reaction environment by means of DTA techniques, and calcination process, activation process of catalysts and the role of promoter were considered from the new point of view of thermal analysis.

Thermal analysis was carried out in the flowing atmosphere of air, SO_2 and $SO_2(7\%)$-air mixture. In the atmosphere of SO_2-air mixture, SO_2 is oxidized to SO_3 by the catalyst so that the atmosphere within the catalyst bed changes to SO_3-SO_2-O_2-N_2 mixture. A gas flow rate of 50-80 ml/min at room temperature and heating rate of 5°C/min were selected as the standard condition.

Catalyst powder sample (100-150 mesh) was prepared by drying the paste-like mixture of the aqueous solution of NH_4VO_3, alkali salt and diatomaceous earth. The drying condition of paste was 110°C and 10 hrs. Alkali salts used were KOH, $KHSO_4$, $K_2S_2O_7$, K_2SO_4, Cs_2SO_4, Na_2SO_4, Li_2SO_4, etc. The mixing ratio of alkali metal to vanadium, M/V, was 2.7 in mole, and that of V_2O_5 to diatomaceous earth was 1:10 in weight. When the $SO_2(7\%)$-air mixture was used as a flowing atmosphere, the catalyst powder calcined up to 600-700°C in air at the heating rate of 5°C/min was used as a sample. By this preliminary treatment, NH_4VO_3 in catalysts completely decomposed to V_2O_5.

Fig.7 shows DTA curves of V_2O_5 catalyst with K_2SO_4 in various atmospheres as an example of M_2SO_4-$V_2O_5(NH_4VO_3)$ systems (M = Li, Na, K, Cs). Curve a shows DTA result for K_2SO_4 alone; the endothermic peak at around 590°C corresponds to the transition of K_2SO_4. Curve b shows the calcination process of catalyst in air; the endothermic peak at around 220°C corresponds to decomposition of NH_4VO_3. Curve c is the repetition of run b, and the endothermic peak at around 450°C corresponds to the melting of the eutectic mixture of K_2SO_4-V_2O_5. Curve d shows DTA curve in flowing SO_2, and curve e is the repetition of run d. Curve f shows DTA curve in static air for the same sample as run e. Curve g shows the result in the flowing $SO_2(7\%)$-air mixture for the catalyst after the calcination process (run b) up to 700°C in air. The exothermic peak at 407°C corresponds to the activation process through which K_2SO_4-V_2O_5 catalyst reacts with SO_3. A broad exothermic peak which appeared just after the activation process corresponds to the catalytic oxidation of SO_2 to SO_3. Curve h is the repetition of run g. The exothermic peak corresponding to activation process disappeared, and only the exothermic peak of catalytic oxidation, reaction process, was observed at around 400-600°C. Curve i is the DTA result in air of the sample after run h. The endothermic peak at 431°C corresponds to the melting of the active complex, K_2SO_4-SO_3-V_2O_{5-x}, which was formed by the activation process at around 400°C on DTA curve g. The melting point of this active complex was lower than the eutectic point of K_2SO_4-V_2O_5 system by about 30°C.

Fig.8 shows the DTA curves in air of various M_2SO_4-V_2O_5 catalysts (M/V=2.7) which have been calcined up to 700°C at heating rate of 5°C/min. Endothermic peaks on the DTA curves from b to e correspond to the eutectic phenomena of various catalysts, except for endothermic peak at 580°C on curve b which corresponds to the transition of Li_2SO_4. It was suggested that V_2O_5 with M_2SO_4 promoter produces eutectic mixture at a lower temperature than the melting point of V_2O_5 alone shown on curve a, and the effect on lowering the melting point of catalysts is the following order: Cs_2SO_4 > K_2SO_4 > Na_2SO_4 > Li_2SO_4.

Fig.9 shows DTA curves in air of the activated catalyst which is a system of M_2SO_4-SO_3-V_2O_{5-x}. Melting points of 305, 370, 407, 420, 410, 508°C of the catalysts containing Cs_2SO_4, $KHSO_4$, K_2SO_4, KOH, Na_2SO_4 and Li_2SO_4 were estimated from endothermal deflection, respectively. All melting points of M_2SO_4-SO_3-V_2O_{5-x} shifted to lower temperatures than M_2SO_4-V_2O_5 systems. The melting temperature was estimated from the point of intersection of the tangent drawn at the point of greatest slope on the leading edge of the endothermic peak with the extrapolated line.

Fig.10 shows DTA curves for fresh M_2SO_4-V_2O_5 catalysts in the flowing SO_2(7%)-air mixture. Exothermic peaks for the activation process at around 400°C and for the oxidation of SO_2 to SO_3 at around 500-600°C appeared on all these curves except for Li_2SO_4-V_2O_5 system. For Cs_2SO_4-V_2O_5 catalyst, furthermore, a small endothermic peak appeared at 395°C just before the exothermic peak corresponding the activating process. This shows that the melting of catalyst occurs first, followed by the activating process.

Fig.11 shows the DTA curves of the activated catalysts obtained by the experiments shown in Fig.10, and the same procedures as Fig.10 were used for all the experiments. The exothermic peak corresponding to the activation process disappeared, and only the exothermic peak for the catalytic oxidation of SO_2 to SO_3 remained. This exothermic peak corresponding to the reaction process always appeared on the DTA curves upon repetition of heating and cooling the system with a good reproducibility.

Catalytic activity of M_2SO_4-SO_3-V_2O_{5-x} systems estimated by the temperature and the area of exothermic peak corresponding to the reaction process gave the following order depending on kind of promoter: Cs_2SO_4 > K_2SO_4 > Na_2SO_4 > Li_2SO_4. Li_2SO_4 has a somewhat negative effect. Diatomaceous earth used as a carrier has no effect on the DTA curves.

In addition, the stability of catalytic activity was tested through the observation of the change of DTA curves with time of holding at 600°C in SO_2(7%)-air atmosphere. The result obtained for KOH-V_2O_5 catalyst is shown in Fig.12. Curve a shows the first DTA run and this process corresponds to activation process. Curve b is the DTA result for the sample heating up to 600°C after the run a. The exothermic peak corresponding to the activation process disappeared. Curve c is the DTA result of the sample treated for 10 hrs, at 600°C after the run b. In this manner, curve c, d, e and f are obtained for the sample kept at 600°C for 10, 15, 20 and 23 hrs, respectively. These times are the sum of total treating time of the sample. It may be suggested from the shape and position of the DTA curves that catalysts of run c to f have stable activity.

Thermal decomposition of $KClO_4$ by Fe_2O_3 catalysts (4)

The catalytic thermal decomposition of $KClO_4$ by Fe_2O_3 catalysts with different preparing history was investigated. Fe_2O_3 was prepared by isothermal calcination of ferrous salts at different temperatures. DTA sample of total weight of 1 g was prepared by mixing $KClO_4$ (-200 mesh) and Fe_2O_3 (-300 mesh) with α-Al_2O_3 in an agate mortar. α-Al_2O_3 was used as a diluent to prevent the melt from falling in drops through the quartz wool layer.

No significant influences of the flow rate of atmospheres

on the DTA peak temperature and shape were observed. From these results, it is suggested that rate determing step of thermal decomposition of $KClO_4$ is not on the diffusion process of a gaseous product such as oxygen from the outer surface to bulk stream. DTA curves of $KClO_4$ recrystallized zero-, three- and five thimes did not differ significantly. Two sieve fractions of $KClO_4$ with particle size of 150-200 mesh and under 200 mesh did not show any change in the DTA peak temperature. On the basis of the preliminary experiments described above, all the experiments were carried out by employing the recrystallized $KClO_4$ (-200 mesh), Fe_2O_3 (-300 mesh) with various preparing histories, and fixed mixing ratio, $KClO_4:Fe_2O_3:Al_2O_3=4:1:5$ in weight.

Fe_2O_3 catalysts used were prepared by isothermal pyrolysis for 1 hr at various temperatures above 700°C for $FeSO_4 \cdot 7H_2O$, and above 500°C for $FeC_2O_4 \cdot 2H_2O$. These catalysts gave X-ray diffraction pattern and IR spectra of pure α-Fe_2O_3.

Fig.13-A shows DTA curves of $KClO_4$ with and without Fe_2O_3 catalysts prepared from $FeSO_4 \cdot 7H_2O$ by pyrolysis in air at 700, 800 and 900°C. Fig.13-B and C also show the DTA curves of $KClO_4$ with and without Fe_2O_3 prepared from $FeC_2O_4 \cdot 2H_2O$ by pyrolysis in air and oxygen, respectively. Five temperatures, 500, 600, 700, 800 and 900°C, were used for the pyrolysis.

The first endothermic peak around 300°C corresponds to the solid phase transition of $KClO_4$ from the rhombic to the cubic. The second small endothermic peak corresponds to fusion of $KClO_4$ (T_2), followed by the exothermic peak of the decomposition (T_3).

It is found in Fig.13 that (a) both T_2 and T_3 shifted to the lower temperature site in all of cases with decreasing the preparation temperature of Fe_2O_3, (b) the addition of catalyst results in the exothermic deflection in low temperature range, before appearance of endothermic peak of fusion of $KClO_4$, and (c) the area of this exothermic peak seems to increase with decreasing preparation temperature of Fe_2O_3. The fact (b) mentioned above is considered to show that the addition of catalyst leads to the solid phase decomposition before fusion of $KClO_4$.

The pronounced catalytic effect is the change in the initial solid phase decomposition temperature (T_1) at which exothermal deflection begins: the lower the preparation temperature of catalysts, the lower the starting temperature of T_1 before the endothermal peak of fusion.

If temperature T_1 may be defined as the characteristic value indicating the catalytic activity of Fe_2O_3, it could be suggested that the catalytic activity in the range of low preparation temperature below 800°C decreases approximately in the order of $FeSO_4$ in air > FeC_2O_4 in O_2 > FeC_2O_4 in air, and increasing with decrease of temperature of preparation.

Effect of additives on solid state reactions

The gas-flow DTA technique was also applied to the following studies.

The effects of $KClO_3$ additive (0-15 wt%) on the thermal decomposition of oxalates, $MC_2O_4 \cdot nH_2O$ (M = Na, K, Mg, Ca, Fe, Ni, Cu, Zn), were studied in three types of flowing atmosphere, N_2, O_2 and air (5,6).

The effects of carbon additive on the chlorination of Mg-containing complex ores (olivine, protoenstatite and talc) with Cl_2-gas (30-50 ml/min) in a temperature range of 25-1000°C were studied (7).

200

APPLICATION OF HIGH-PRESSURE DTA APPARATUS

High-pressure DTA apparatus (8)

Fig.14 shows the main part and the attached measuring instruments. The main part consists of a twin type autoclave with identical chembers for reaction and reference (60 ml capacity), embedded in an aluminum heater block. This apparatus was applicable up to 200 kg/cm² and 500°C. In the center of each chember, pressure-proof tubes are inserted, and Chromel-Alumel thermocouples (d=0.3mm) were inserted into each of these tubes to enable to measure differential temperature and the reference chamber temperature. Further, both chambers are connected to pressure transducers. This apparatus is available to measure simultaneously the differential temperature (DTA curve) and the differential pressure (DPA curve) under reaction automatically.

In order to facilitate the gas-liquid contacts, the liquid phase was agitated by an electromagnetic stirrer set at the bottom of the reaction chamber. The samples in the reaction chamber were agitated by a rotating magnetic piece within the reaction chamber.

Identical micro heating coils were placed at the tips of the pressure-proof tubes inserted in both chambers. By sending a limited electric current to the coil in the reaction chamber, the relationship between the given electric calories and the resulting thermal areas was measured. Linear relationship between the two was obtained in the range of temperature between 25 and 200°C. Reaction heat of 593 cal/g-benzene for the benzene hydrogenation was calculated using this relationship. This value well agreed with that in the literature.

In spite of using the thick walls for these chambers (od.= 50mm, id.=25mm), a very good reproducibility of peak temperature (within ±1.5°C on repeated runs) was obtained under the controlled heating rate and stirring. The sensitivity was such that electrically generated heat of 2 cal/min could be measured at a heating rate of 0.7°C/min.

High pressure hydrogenation of aromatic compounds (9,10)

High-pressure DTA apparatus was applied to the studies of high pressure hydrogenation and hydrogenolysis of phenol, benzene and diphenylether using six kinds of commercial catalysts: (A)= stabilized nickel (50% Ni-diatomaceous), (B)=sulfur-proof nickel (45% Ni-copper-chromium oxide-diatomaceous), (C) and (D)=copper-chromium oxide (modified Adkins type catalysts), (E)=5% palladium-active carbon, (F)=5% platinum-activated carbon.

At an initial hydrogen pressure of 100 kg/cm², both DTA and DPA curves were measured simultaneously up to 450°C. Three grams sample with 0.3 g catalyst and 3 g α-Al₂O₃ as a reference material were used. Rotation of electromagnetic stirrer in the sample chamber was 600 r.p.m. and the heating rate was 2°C/min. Sample was taken out of the apparatus at various points on the DTA curve and products were analysed by means of IR and gas chromatographic methods.

For an example, DTA and DPA curves for high pressure hydrogenation of phenol with various catalysts are shown in Fig.15. From these curves and analysis of products during the reaction process, the differences in reaction course and a comparison of

each catalyst activity were investigated. The reaction course
with catalysts (A) and (B) was as follows: phenol → cyclohexanol
→ cyclohexane → methane. However, for catalysts (C), (D), (E)
and (F), the reaction of cyclohexane → methane did not proceed
even at 500°C. For catalyst (E), cyclohexanone was clearly observed to be intermediate during the reaction of phenol to cyclohexanol. The order of the catalytic activities in the hydrogenation reaction from phenol to cyclohexanol was as follows: (E)=
(F) > (B) > (A) > (C) > (D).
        Kinetic parameters were also estimated by using Kissinger's
method. As an example, Table 1 shows a comparison between kinetic data from DTA and isothermal batch method for liquid phase
hydrogenation of benzene. Approximately the same kinetic results
were obtained. Similar kinetic results were obtained for both
revolution speeds, 300 and 800 r.p.m. This means that the diffusion process of hydrogen does not control the over-all reaction
rate.

Activity test of coal-hydrogenation
catalysts under high pressure (11)

        High-pressure DTA technique was applied for coal hydrogen-
ation with various catalysts under 200 atmospheres, and activities
of catalysts were compared from DTA peak. Sumiyoshi coal-Japan
was used as a sample coal. Water-insoluble catalysts (-170 mesh)
were mixed mechanically with coal powder (-200 mesh). Water-
soluble catalysts were combined by evaporating an aqueous solution
containing suspended solid coal powder. Five grams of sample with
catalyst and 7 g of α-Al_2O_3 as a reference material were used.
Initial hydrogen pressure was 100-120 kg/cm^2 and heating rate was
2°C/min.
        Fig.16 shows DTA curves for high pressure hydrogenation of
coal with Fe_2O_3 and red mud catalyst (Fe_2O_3:41.5, Al_2O_3:21.4,
SiO_2:12.4 in wt%). Curve No.24 is that of coal without catalyst.
A broad exothermic peak at 420-440°C corresponds to the coal
hydrogenation. Curve No.34 is the result for coal (4.5g) with
red mud (0.5g), and similar to No.24. No catalytic activity of
red mud was shown. When elemental sulfur is added to red mud
catalyst, however, a steep exothermic peak appears at a lower
temperature than for coal with red mud. The role of elementary
sulfur on promoting the activity of red mud catalyst will be
described later.
        Fig.17 shows DTA curves for the coal hydrogenation with
$SnCl_2 \cdot 2H_2O$, SnO_2 and SnS catalysts, respectively. It was found
that catalytic activity differs from one another depending upon
the kind of salt. Fig.18 shows the DTA curves for the coal
hydrogenation with $ZnCl_2$, ZnO and ZnS catalysts. The highest
activity was shown by $ZnCl_2$ catalyst.
        The order of the catalytic activity estimated from DTA
exothermic peak corresponding to coal hydrogenation reaction
was as follows: $ZnCl_2$ ($ZnCl_2$, 329°C) > $SnCl_2 \cdot 2H_2O$ (?, 339°C) >
SnS (SnS+Sn, 372°C) > SnO_2 (SnO_2+Sn, 386°C) > $(NH_4)_6Mo_7O_{27}$ (MoO_2,
390°C) > red mud + S (?, 397°C) > ZnO (ZnO, 401°C) > red mud (?,
429°C) > ZnS (ZnS, 420-435°C) = no catalyst (420-436°C). Metallic compounds and temperature in the parentheses show the
chemical forms of catalysts under the reaction conditions and
the exothermic peak temperature, respectively. In these experiments, $ZnCl_2$ was found to be the most active catalyst, and both

red mud and ZnS have no catalytic effect.

One of the major problems is that the coal hydrogenation is a very complex one in which many elemental reactions overlap. At the present time, it is difficult to analyse the details of each elemental reaction in DTA exothermic peak which was described as one corresponding to the coal hydrogenation reaction. Also different kind of reaction could be caused by using a different kind of catalyst. Therefore, it must be noted that a direct comparison of catalytic activity from DTA peak to such a complex reaction is not comparison of a single elemental reaction but that of the DTA peak which was composed of these many elemental reactions.

Iron Oxide-sulfur catalysts for the coal hydrogenation (12)

The promoting ability of elemental sulfur for the high pressure coal hydrogenation with red mud and iron oxide catalysts was studied. In Fig.16, from a comparison of the shape and position of DTA exothermic peaks (at 350-450°C) between the catalysts with and without sulfur, the promoting capability of sulfur is clear. Furthermore, when sulfur was added in the system, another exothermic peak appeared at 200-250°C. It may be inferred that this peak newly generated by the reaction between iron oxide and sulfur, and active catalyst was resulted.

Fig.19 shows the DTA curves of Fe_2O_3 with sulfur and red mud with sulfur under high pressures of hydrogen, nitrogen and helium. For the Fe_2O_3-S system in hydrogen, a large exothermic peak at about 220°C and an endothermic peak at 300°C appeared. X-ray analysis of sample taken out of the apparatus at about 500°C showed it to be FeS. No Fe_2O_3, Fe_3O_4, FeO and Fe were detected. The samples at about 250°C just after the large exothermic peak were mixtures of FeS_2 and FeS. An endothermic peak at about 300°C probably corresponds to the reaction of FeS_2 to FeS.

Red mud catalyst with sulfur showed a similar thermal behavior to Fe_2O_3, though the location of the peak shifted slightly. The final product was FeS as with the Fe_2O_3 catalyst.

In N_2 and He atmospheres, Fe_2O_3-S system had an endothermic peak corresponding to the melting of sulfur at 100-150°C and an exothermic peak at about 450°C. The final product was FeS_2. Red mud-S system had an exothermic peak at about 280°C but no peak appeared about 450°C. The final products were Fe_2O_3 and FeS_2.

From such a thermal analysis of the catalytic hydrogenation reaction and of these catalysts of iron oxide-S systems under high pressure of H_2, N_2 and He, and from the X-ray diffraction analysis of catalyst sample through the reaction course, the following conclusions were reached: 1) The red-mud showed a similar thermal behavior to Fe_2O_3. 2) Red mud and Fe_2O_3 showed no catalytic effect but their catalytic activities were promoted by the addition of elementary sulfur. 3) Fe_2O_3 changes to FeS in the course of the coal hydrogenation reaction under a high pressure of hydrogen, when the catalytic activity appears at about 350-400°C. 4) The following reaction scheme was suggested, a) melting of S, endothermic at 110-140°C; b) $S + H_2 \rightarrow H_2S$, in the presence of Fe_2O_3; c) $Fe_2O_3 + H_2S \rightarrow FeS + FeS_2 + H_2O$, exothermic at 200-.240°C, partly $Fe_3O_4 + H_2S \rightarrow FeS + FeS_2 + H_2O$; d) $FeS_2 + H_2 \rightarrow FeS + H_2S$, endothermic at 270-330°C.

Fig.20 shows the effect of the ratio S/red-mud on catalytic activities of red mud for high pressure hydrogenation of coal.

Red mud of 10 wt% to coal was used. The exothermic peak at 350-400°C corresponding to the hydrogenation reaction shifted to the lower temperature site, and catalytic activity increases with increasing the ratio S/red-mud at the time of preparation. The exothermic peak at 200-250°C corresponding to the activating reaction of $Fe_2O_3 + S + H_2 \rightarrow FeS + FeS_2 + H_2O$ becomes progressively greater as the ratio S/red-mud increases.

Fig.21 shows the DTA curves for the coal hydrogenation by two FeS catalysts. The first is the synthesized FeS which was prepared by heating Fe_2O_3-S mixtures up to 500°C at heating rate of 2°C/min (run No.82 in Fig.19); the second is commercial FeS reagent. The synthesized FeS clearly shows catalytic activity, but the commercial one small activity. It seems to be due to preparative conditions, e.g. the ratio of S to Fe. The synthesized FeS was prepared from the ratio of 3.1 g of S to 3.9 g Fe_2O_3, while the commercial one has excess Fe.

CONCLUSIONS

The simple gas-flow DTA and the high-pressure DTA apparatus designed were found to be useful to follow the various catalytic reaction courses and the catalytic actions under working state of the catalysts, and to compare the catalytic activities. Further use of thermoanalytical techniques for the study of new catalytic materials can be expected.

ACKNOWLEDGEMENT

The author wishes to thank Prof. G. Takeya, Hokkaido Univ., for helpful discussions and for the design of high-pressure DTA apparatus.

REFERENCES

1. T.Ishii, T.Furumai and G.Takeya, J.Chem.Soc.Japan (Ind.Chem Sec.) 70, 1652 (1967)
2. S.Shimada, R.Furuichi and T.Ishii, J.Chem.Soc.Japan (Ind. Chem.Sec.) 74, 2006 (1971)
3. T.Ishii, M.Aramata and R.Furuichi, J.Chem.Soc.Japan (Chem. Ind.Chem.) 266, 1068 (1972)
4. R.Furuichi, T.Ishii and K.Kobayashi, J.Therm.Anal. 6, 305 (1974)
5. T.Ishii, K.Kamada and R.Furuichi, J.Chem.Soc.Japan (Ind. Chem.Sec.) 74, 854 (1971)
6. T.Ishii, R.Furuichi, T.Kawasaki and K.Kamada, Bull.Fac.Eng. Hokkaido Univ. 67, 137 (1973)
    T.Ishii, R.Furuichi and C.Okawa, ibid. 71, 163 (1974)
7. T.Ishii, R.Furuichi and Y.Kobayashi, Thermochimica Acta, 9, 39 (1974)
8. G.Takeya, T.Ishii, K.Makino and S.Ueda, J.Chem.Soc.Japan (Ind.Chem.Sec.) 69, 1654 (1966)
9. T.Ishii, T.Yahata and G.Takeya, Chem.Eng.Japan, 31, 896 (1967
10. S.Ueda, S.Yokoyama, T.Ishii and G.Takeya, J.Chem.Soc.Japan (Ind.Chem.Sec.) 74, 1377 (1971)
11. T.Ishii, Y.Sanada and G.Takeya, J.Chem.Soc.Japan (Ind.Chem. Sec.) 71, 1783 (1968)
12. T.Ishii, Y.Sanada and G.Takeya, J.Chem.Soc.Japan (Ind.Chem. Sec.) 72, 1269 (1969)

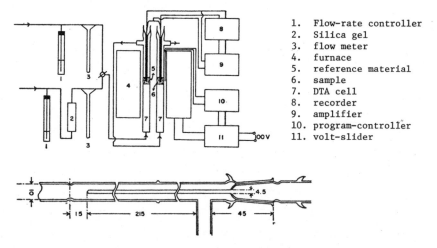

1.	Flow-rate controller
2.	Silica gel
3.	flow meter
4.	furnace
5.	reference material
6.	sample
7.	DTA cell
8.	recorder
9.	amplifier
10.	program-controller
11.	volt-slider

Fig. 1-A,B. Schematic diagram of gas-flow DTA apparatus and cell. Unit of dimension : mm

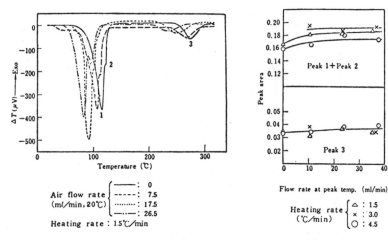

Fig.2. DTA curves of dehydration of $CuSO_4 \cdot 5H_2O$ by flowing gas technique.

Fig.3. Effect of air flow rate on peak area for dehydration of $CuSO_4 \cdot 5H_2O$.

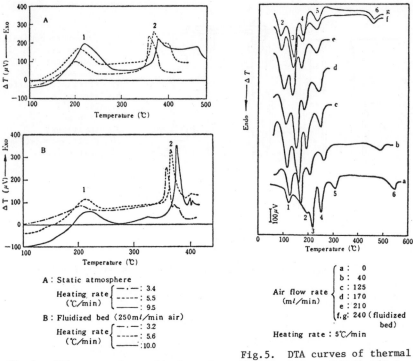

A : Static atmosphere

Heating rate
(℃/min) { —·— : 3.4 / ----- : 5.5 / —— : 9.5

B : Fluidized bed (250ml/min air)

Heating rate
(℃/min) { —·— : 3.2 / ----- : 5.6 / —— :10.0

Fig.4. DTA curves for oxidation of UO₂ in air.

Air flow rate
(ml/min) { a : 0 / b : 40 / c : 125 / d : 170 / e : 210 / f,g: 240 (fluidized bed)

Heating rate : 5℃/min

Fig.5. DTA curves of thermal decomposition of MgCl₂·6H₂O by air flowing technique.

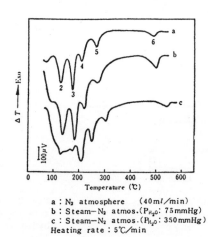

a : N₂ atmosphere (40ml/min)
b : Steam—N₂ atmos.(P$_{H_2O}$: 75mmHg)
c : Steam—N₂ atmos.(P$_{H_2O}$: 350mmHg)
Heating rate : 5℃/min

Fig.6. DTA curves of thermal decomposition of MgCl₂·6H₂O in steam.

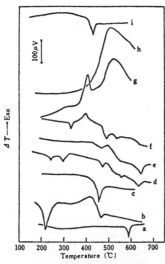

a : K₂SO₄+α-Al₂O₃ (air, static)
b : K₂SO₄−V₂O₅ (air), "calcination process"
c : repetition of b (air, static)
d : K₂SO₄−V₂O₅ (SO₂ flow)
e : repetition of d (SO₂ flow)
f : same sample as e (air, static)
g : K₂SO₄−V₂O₅ catalyst after calcination process (SO₂+air, flow), "activation process"
h : repetition of g (SO₂+air, flow), "reaction process"
i : repetition of h (air, static)

Fig.7. DTA curves of K₂SO₄−V₂O₅ in various atmospheres. ( ) shows atmosphere.

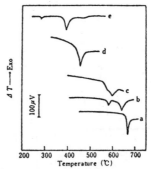

a : V₂O₅+diatomaceous earth
b : Li₂SO₄−V₂O₅ catalyst
c : Na₂SO₄−V₂O₅ catalyst
d : K₂SO₄−V₂O₅ catalyst
e : Cs₂SO₄−V₂O₅ catalyst

Fig.8. DTA curves of catalysts after calcination process (atmosphere: air, heating rate:5°C/min).

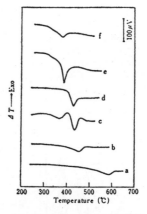

a : Li₂SO₄−V₂O₅ catalyst (508℃)
b : Na₂SO₄−V₂O₅ catalyst (410℃)
c : KOH−V₂O₅ catalyst (420℃)
d : K₂SO₄−V₂O₅ catalyst (407℃)
e : KHSO₄−V₂O₅ catalyst (370℃)
f : Cs₂SO₄−V₂O₅ catalyst (305℃)

Fig.9. DTA curves (melting point) of activated catalyst (atmosphere: air, heating rate:5°C/min).

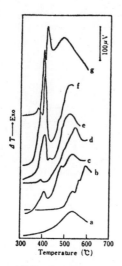

a : V₂O₅+ diatomaceous earth
b : Li₂SO₄—V₂O₅ catalyst
c : Na₂SO₄—V₂O₅ catalyst
d : KHSO₄—V₂O₅ catalyst
e : K₂SO₄—V₂O₅ catalyst
f : KOH—V₂O₅ catalyst
g : Cs₂SO₄—V₂O₅ catalyst

Fig.10.   DTA curves of
catalysts (atmosphere:
SO_2+air, flow, heating
rate:5°C/min).
ACTIVATION PROCESS.

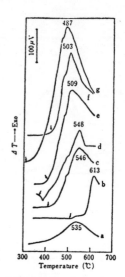

a : V₂O₅+ diatomaceous earth
b : Li₂SO₄—V₂O₅ catalyst
c : Na₂SO₄—V₂O₅ catalyst
d : KHSO₄—V₂O₅ catalyst
e : K₂SO₄—V₂O₅ catalyst
f : KOH—V₂O₅ catalyst
g : Cs₂SO₄—V₂O₅ catalyst
(↓ is melting point of activated
catalysts, see Fig. 9)

Fig.11.   DTA curves of
activated catalysts
(atmosphere:SO_2+air,
flow, heating rate:5°C/
min).
REACTION PROCESS.

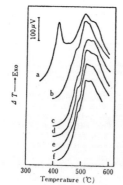

time kept
at 600℃

a :   0 hr "activation process"
b :   0 hr "reaction process"
c : 10 hr
d : 15 hr
e : 20 hr
f : 23 hr

Fig.12.   Change of DTA curves of
KOH–V₂O₅ catalyst with time kept
at 600°C.

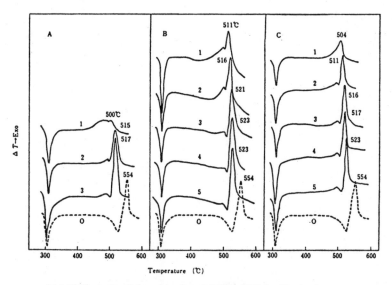

A : α – Fe₂O₃ prepared by calcination of FeSO₄·7H₂O for 1 hr. in air.
   Calcination temp. (℃) 1:700, 2:800, 3:900
B : α – Fe₂O₃ prepared by calcination of FeC₂O₄·2H₂O for 1 hr. in air.
   Calcination temp. (℃) 1:500, 2:600, 3:700, 4:800, 5:900
C : α – Fe₂O₃ prepared by calcination of FeC₂O₄·2H₂O for 1 hr. in O₂.
   Calcination temp. (℃) 1:500, 2:600, 3:700, 4:800, 5:900
O : KClO₄ alone (0.04 g)

Fig.13. Catalytic activities of α-Fe₂O₃ for decomposition of
KClO₄. KClO₄(0.32g) – α-Fe₂O₃(0.08g). Heating rate:5°C/min.

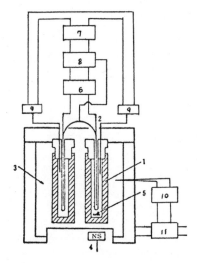

1.   autoclave reactor
2.   thermocouple
3.   aluminum block and heater
4.   stirrer
5.   magnetic rotator
6.7. microvolt amplifiers
8.   recorder
9.   pressure transducers
10. program controller
11. transformer

Fig.14. High-pressure DTA apparatus

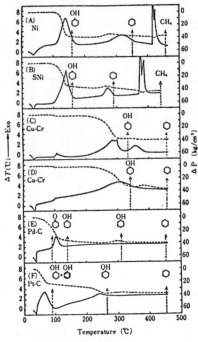

── : DTA curves
········ : DPA curves
Heating rate: 1.6℃/min
Initial H_2 pressure: 100 kg/cm² (20℃)

Fig.15. DTA and DPA curves for high pressure hydrogenation of phenol.

Table 1 Kinetic data from differential thermal analysis and isothermal batch methods for liquid phase hydrogenation of benzene

		DTA method	Isothermal method
E [kcal/mol]		20.0	18.3
ln A		21.9	18.6
n		1.1	1.0
k_r [min⁻¹]	124℃	0.0310	0.0236
	139℃	0.0778	0.0557
	150℃	0.1470	0.0983

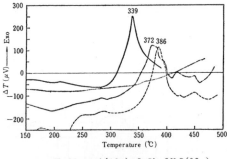

No.24　coal alone (5.0 g)
No.79　coal (4.8 g) + Fe₂O₃ (0.2 g)
No.78　coal (4.6) + Fe₂O₃ (0.2) + S (0.16)
No.34　coal (4.5) + red mud (0.5)
No.30　coal (4.5) + red mud (0.5) + S (0.05)
Heating rate: 2℃/min
Reaction H_2 pressure: 200〜230 kg/cm²

Fig.16. DTA curves for high pressure hydrogenation of coal by Fe₂O₃ and red mud catalysts.

── : No.36　coal (4.2 g) + SnCl₂·2H₂O (0.8 g)
········ : No.41　coal (4.2) + SnCl₂·2H₂O (0.8)
　　　　　　[He atmosphere]
──── : No.47　coal (4.4) + SnO₂ (0.6)
─·─ : No.46　coal (4.4) + SnS (0.6)
Heating rate : 2℃/min
Reaction H_2 pressure : 190〜210 kg/cm²

Fig.17. DTA curves for high pressure hydrogenation of coal by Sn catalysts.

210

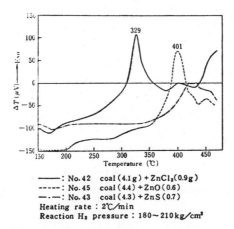

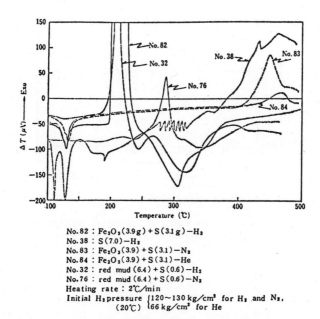

- : No.42 coal (4.1g) + ZnCl₂ (0.9g)
---- : No.45 coal (4.4) + ZnO (0.6)
-·- : No.43 coal (4.3) + ZnS (0.7)
Heating rate : 2℃/min
Reaction H₂ pressure : 180~210kg/cm²

Fig.18. DTA curves for high pressure hydrogenation of coal by Zn catalysts.

No.82 : Fe₂O₃ (3.9g) + S (3.1g) −H₂
No.38 : S (7.0) −H₂
No.83 : Fe₂O₃ (3.9) + S (3.1) −N₂
No.84 : Fe₂O₃ (3.9) + S (3.1) −He
No.32 : red mud (6.4) + S (0.6) −H₂
No.76 : red mud (6.4) + S (0.6) −N₂
Heating rate : 2℃/min
Initial H₂ pressure ⎰120~130 kg/cm² for H₂ and N₂,
        (20℃) ⎱66 kg/cm² for He

Fig.19. DTA curves of Fe₂O₃ + S and red mud + S systems under pressure of H₂, N₂ and He.

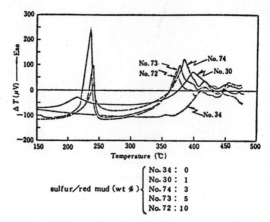

sulfur/red mud (wt %) $\begin{cases} \text{No. 34 :} & 0 \\ \text{No. 30 :} & 1 \\ \text{No. 74 :} & 3 \\ \text{No. 73 :} & 5 \\ \text{No. 72 :} & 10 \end{cases}$

Fig. 20. Effect of adding sulfur on catalytic activities of red mud for high pressure hydrogenation of coal.

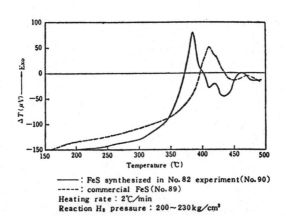

——— : FeS synthesized in No. 82 experiment (No. 90)
----- : commercial FeS (No. 89)
Heating rate : 2°C/min
Reaction H_2 pressure : 200~230 kg/cm²

Fig. 21. DTA curves for high pressure hydrogenation of coal by FeS catalysts.

212

APPLICATION OF THERMAL ANALYSIS TO NEW POLYMER SYSTEMS

J. J. Maurer
Corporate Research Laboratories
Exxon Research and Engineering Company

ABSTRACT

Thermal analytical techniques have extensive utility in both the research and applied areas of polymer science. This review describes specific applications of important thermal analytical techniques to selected polymer systems. These systems, which cover a spectrum of polymer types and compositions, include block polymers, ionomers and polymer blends. Much of the work relates to the general area of structure-property relationships in polymers.

INTRODUCTION

The polymer field offers an unusually large opportunity for the generation of new materials due to advances which have taken place in polymer synthesis techniques as well as the opportunity for property optimization via polymer blends or composites. Experience has shown that there are many opportunities for the use of thermal analytical techniques in this complex field. The objectives of this discussion are to survey selected recent activities in several of the key areas of polymer research, illustrate the kinds of thermal techniques which are being used and the information they provide and, hopefully, stimulate discussion regarding data interpretation, opportunities for new or alternative techniques, etc. The major areas which will be considered in this report are the following:

I. Illustration of polymer transition characteristics
II. Analysis and control of sequence distribution
III. Ionomers
IV. Block polymers
V. Compatible polymer blends
VI. Problems encountered during thermal analysis of new materials.

DISCUSSION

Illustration of Polymer Transition Characteristics

An extensive discussion of the transition behavior of polypentenamer has been given by Wilkes and co-workers.[1] Consideration of the transition characteristics of this polymer is a useful way to begin a discussion of the application of thermal analytical (TA) methods to polymer systems. For our present purposes, the following aspects of this work are of interest (see Figures 1 and 2).

1. The type of thermogram characteristics associated with the glass transition (Tg), polymer crystallization (Tcr) and polymer melting (Tm) are shown at about -92°, -67° and +13°C, respectively. Note the baseline shift due to the heat capacity change at Tg, the crystallization exotherm and the endotherm corresponding to polymer melting.

213

2. As illustrated here, variations in degree of crystallization and melting as well as in Tm, are frequently observed, and studied, by control of thermal history of the sample.

3. The effect of polymer structure (cis/trans ratio) on Tg, Tcr and Tm is indicated.

4. As shown in Figure 2, measurement of Tg or Tcr + Tm would enable determination of polymer composition (trans content) for this system.

Analysis and Control of Sequence Distribution

An important feature of many polymeric systems is the existence of variability of chain microstructure or sequence distribution. Depending upon the chemical characteristics of the monomers, their relative reactivity, and the polymerization system (including catalyst) a wide variety of microstructures are possible. These range from crystalline to amorphous. Further, the latter category can vary from perfectly alternating, to random, to a more complex case where the sequence lengths of one or both monomer units in the chain are sufficient (deliberately or accidentally) to permit one or more types of association between these block structures (either on a micro or macro scale). These microstructure variations, depending on type and degree, can have a major influence on the bulk, melt or solution properties of a polymer, and thus on its potential utility for a given practical application. Learning to control polymer microstructure is therefore an important step toward developing new polymer systems. The following discussion is concerned with two of the newer developments in the microstructure area. One relates to achieving an improved understanding of the Tg value for copolymers; the other is concerned with control of Tg by synthesis conditions.

A. Analysis of Copolymer Glass Transition Characteristics

The relationship between copolymer composition and Tg is commonly analyzed via the following equations which are due to Fox (Eq. 1) or Gordon-Taylor-Wood (Eq.2).

$$\frac{1}{Tg_P} = (\frac{W_A}{Tg_A}) + (\frac{W_B}{Tg_B}) \tag{1}$$

Here, Tg_P is the glass transition temperature for a copolymer which contains weight fractions W_A and W_B of the two monomer units, and the homopolymers of these units have glass transition values of Tg_A and Tg_B respectively.

$$Tg_P = [Tg_A + (KTg_B - Tg_A)W_A]/[1 - (1 - K)W_B] \tag{2}$$

Here, K is a constant for the given copolymer and is reportedly related to the specific volumes of homopolymers A and B at their respective Tg's.

It has been found that some copolymer systems are well described by these equations, i.e. linear relationships are observed when Tg is plotted vs the appropriate function of polymer composition. Many other cases are known, however, where concave or convex relationships are encountered. Treatment of such data by linear least-squares techniques often predicted

214

homopolymer Tg values which were in major disagreement with experimentally determined values. Interpretation of these systems has required some new insight into the matter, such as the following work by Johnston.[2]

The basic point of this treatment is that the common linear relationships for Tg, by assuming that a given monomer unit will contribute the same free volume to a copolymer as it does in a homopolymer neglect the possibility of steric and energetic effects due to adjacent dissimilar units in the chain. His approach is to take into account the sequence distribution in the polymer, and particularly to assign Tg values to AB dyads and other sequences if required. Here reference is made to the fact that the environment of a given monomer unit in the chain, e.g. monomer A, can be AA, AB or BB. An equation was developed which takes into consideration the copolymer sequence distribution and AA, AB and BB Tg's. The required probabilities of these linkages (P_{AA}, P_{AB}, P_{BB}), were calculated from the monomer feed and the copolymer reactivity ratios.

$$\frac{1}{Tg_P} = \frac{W_A P_{AA}}{Tg_{AA}} + \frac{W_A P_{AB} + W_B P_{BA}}{Tg_{AB}} + \frac{W_B P_{BB}}{Tg_{BB}} \tag{3}$$

For many cases, use of a Tg_{AB} for AB dyads was a sufficient correction to enable prediction of sequence effects in a copolymer series via Eq. 3. In other systems, or in the case of terpolymers, there is a strong triad-Tg effect and an expanded Eq. 4 was required to accomodate these polymer characteristics. However, Eq. 3 was adequate for prediction of Tg values for the AN/MMA copolymer series.

$$\frac{1}{Tg_P} = \frac{W_A P_{AA}}{Tg_{AA}} + \frac{W_B P_{BB}}{Tg_{BB}} + \frac{W_C P_{CC}}{Tg_{CC}} + \frac{W_A P_{AB} + W_B P_{BA}}{Tg_{AB}}$$
$$+ \frac{W_A P_{AC} + W_C P_{CA}}{Tg_{AC}} + \frac{W_B P_{BC} + W_C P_{CB}}{Tg_{BC}} \tag{4}$$

This approach can be illustrated for AN/MMA copolymers. Tg values were measured by DSC and Tg_{AB} was obtained by a computer multiple regression program. Tg's of random AN/MMA copolymers were then calculated using Eq. 3, and values of 79°C for Tg_{AB} and 105°C for both PMMA and PAN. Figure 3, which compares the Fox vs sequence-effect treatments of the Tg-composition data for this system, shows that the latter approach was highly successful in this instance. Another system α MS/MMA was found, as expected, to obey the Fox equation. This was attributed to the fact that both monomer units have two substituent groups which are large but non-equivalent; thus, it was reasoned that there is not a large resultant change in steric crowding along the backbone of the polymer.

As indicated previously, analysis of terpolymers is a more challenging case because of the larger number of dyad types present. Thus Tg prediction for an αMS/MMA/AN terpolymer involves the linear behavior of αMS-MMA dyads plus the sequence dependent behavior of the αMS-AN and MMA-AN dyads. Despite the complexity of this system, the predicted Tg values were in close agreement with the measured values as shown in Table 1.

TABLE 1

Tg Analysis of αMS/MMA/AN Terpolymers

Terpolymer, m%			Tg, °C		
αMS	MMA	AN	Expt.	Seq. Dist. (a)	Fox (b)
65.6	20.2	14.2	148	148	155
64.1	13.9	22.0	147	145	156
59.7	12.2	28.1	142	138	155
59.6	22.5	17.9	142	142	151
45.3	23.2	31.5	131	128	143
44.9	23.5	31.6	130	127	143

(a) Predicted using Eq. 4.
(b) Predicted using the terpolymer form of the Fox equation.

Thus, Johnston's work appears to represent a step forward regarding our understanding of the complex relationships in copolymer systems and, hopefully, also in our ability to synthesize polymers with specific glass transition/microstructure relationships.

B. Tg Variation via New Polymerization Techniques

Another important development in the Tg vs microstructure area was the discovery that the sequential structure of copolymers can be regulated when vinyl compounds are copolymerized in the presence of certain metal halides such as alkyl aluminum halides.[3] This makes it possible to compare the properties of random copolymers and alternating polymers of the same composition. Based on this development, an extensive study of Tg characteristics has been made for equimolar copolymers which differed in sequence distribution (alternating vs random).[3] As shown in Figures 4-7, the DSC-determined Tg values are not always in agreement for such pairs of polymers, thus indicating varying degrees of microstructure effect(s). Three types of behavior are illustrated in these figures. It is evident that the Tg of an equimolar alternating copolymer can be greater than, the same, or less than that of the corresponding random copolymer. Since polymer properties vary with Tg value, this work indicates a potential route to certain new materials, and the role that "routine" TA measurements can play in such investigations.

Ionomers

Ionomers are a class of polymers which are capable of forming intermolecular ionic bonds. Two common examples of these materials are prepared by copolymerizing an olefin (such as ethylene or styrene) with a carboxylic acid (such as acrylic acid or methacrylic acid) and subsequently ionizing the copolymer to form ionomers. The properties of the ionomer are markedly different from the parent copolymer as illustrated in Table 2. Some of these properties, which are listed below, are of practical interest while others (creep behavior and permanent set) have proved unacceptably large for some applications. Among the attractive properties which can be observed are increase in optical clarity, impact resistance, tensile strength and viscosity.

TABLE 2

COMPARISON OF PROPERTIES: POLYACRYLIC ACID VS ZINC POLYACRYLATE[a]

| | Property | | |
Polymer	Shear Modulus (dynes/cm^2)	Compression Strength[b] (lbs/in^2)	Coefficient of Expansion (per deg. C)
Polyacrylic Acid[c]	2.78×10^{10}	25,300	5.52
Zinc Poly-acrylate	6.2×10^{10}	46,800	1.44

(a) After Nielsen, ref. 1.
(b) At 25°C
(c) 94 percent acrylic acid and 6 percent 2 ethylhexyl acrylate copolymer.

One of the frequently studied properties of ionomer systems is the glass transition temperature (Tg) which has been treated in a review by Eisenberg.[18] Characteristically, Tg increases as ionic groups are incorporated in the polymer. Two mechanisms have been proposed for this increase: (a) cross-linking by ionic species, and (b) a copolymerization effect on Tg similar to that observed in non-ionic polymers. This area is the subject of continuing research.

An illustration of the glass transition characteristics of selected ionomers is useful to indicate both the basic nature of these materials as well as one means by which thermal analysis is being employed in this active field of polymer structure-property research. Figure 8 indicates the effect of ion concentration on Tg of styrene-sodium methacrylate copolymers.[4] Tg was determined by differential scanning calorimetry using a Perkin-Elmer Model DSC-1. Heating and cooling rates were 10°C/min. This data illustrates the increase in glass transition temperature with increasing ion content which is usually observed for ionomers. The effect of molecular weight is also indicated. The increase in Tg with molecular weight is expected and parallels the general behavior of non-ionic polymer systems.

The complexity of this area of research is illustrated in Figure 9 which is concerned with polymers of n-butyl methacrylic acid and ionomers derived therefrom. Two aspects of the data are of interest: (a) the similarity in behavior of the ionomers and unionized polymers, and (b) the shape of the curves, particularly the apparent maximum or shoulder. This effect has been postulated to involve one or more of the following factors: (a) the plasticizing effect of water molecules, (b) a copolymer composition effect on Tg, and (c) a change in the distribution of ions in the polymer matrix, either the initiation of formation of domains, or a change from an increase in the number of domains to an increase in the size of these microstructures. Williams made use of the type of microstructure variations mentioned elsewhere in this discussion to examine the possibility that block ion formation was involved in the shoulder of Figure 9.[5] Com-

217

parison of random vs perfectly alternating styrene/n-butyl methacrylate copolymers indicated that both followed a relationship similar to that of Figure 9. Therefore, this effect does not appear to be related to block ion formation.

The nature of the ionomer morphology which gives these polymers such unique properties is a matter of controversy and active research. The subject is inherently complex and requires a variety of methods and techniques for analysis as indicated in the following discussion.

One of the classes of ionomers involves copolymers containing ethylene. These systems offer an additional element of complexity because of the ability of the ethylene units to crystallize. Two models have been put forth to explain the morphology of this type of system. The "homogeneous" model assumes that the ionic groups are uniformly distributed in the amorphous phase of the polymer. In contrast, the "cluster" model considers that these ionic groups are present in aggregates of about 100A diameter. This latter model assumes that the morphology consists of amorphous phase, crystalline phase and ionic clusters. Wide and small angle x-ray scattering data have been used to support various views regarding these morphological models.

Cooper et. al.[7] have extensively investigated a series of polymers including polyethylene, ethylene/methacrylic acid, butadiene/methacrylic acid and ionomers derived from these copolymers. Wide and small angle x-ray scattering data were used to examine the possible origin of a well-known peak in the wide angle scattering measurements for polyethlene ionomers that is not observed in the parent copolymers. This peak has been attributed to a change in morphology upon ionization.

Cooper et. al. make use of differential scanning calorimetry (heating rate: 10°/min; sample size: 20.2 mg; DuPont Model 900 Thermal Analyzer) to support the x-ray scattering analyses of these samples. For example, the thermograms of Figure 10 show that, relative to polyethylene, the amount of crystallinity and the melting point are reduced by copolymerization and ionization. This behavior was similar to the variation in low angle x-ray scattering patterns for the same materials. In particular: (a) DSC melting points and Bragg spacings were closely related; (b) the amount of crystallinity (by DSC) and the intensity of small-angle x-ray scattering were related, and (c) small angle x-ray scattering was absent above the melting point of the ethylene ionomer, and in non-crystalline ionomers. These results were interpreted as an indication that the small-angle x-ray scattering is due to polyethylene crystalline lamellae rather than to ionic domains.

This work led to a new "aggregate" model for ionomers which is similar to the "homogeneous" model, in that acid aggregates are assumed to be homogeneously distributed, and also to the "cluster" model in that acid groups are assumed to aggregate. Points of difference in the new model are that the degree of aggregation is assumed to be much lower and influenced by carboxyl group concentration and the presence of any polar diluents. Aggregate size is indicated to increase from dimers to trimers to tetramers up to septimers as the acid content in the copolymer increases, and also to increase via plasticization with water. Figure 11 is a schematic representation of such a model for polyethylene-methacrylic acid copolymer (ionized 1/3). The black dots represent ionomer aggregates.

One additional account of ionomer research will be considered for the following reasons: additional information about ion clustering is presented, the work relates to a styrene-based ionomer, and it illustrates the use of dynamic mechanical analysis-another type of "thermal analysis" which is widely employed, in one form or another, in polymer characterization studies. Previous work by Eisenberg and coworkers[4,8] presented rheological evidence to support a hypothesis that in styrene ionomers ion clustering occurs above 6 mol % ions whereas simple ion multiplets occur below that concentration. Viscoelastic relaxation studies showed that two master curves, each with its own set of shift factors, were needed to achieve a complete description of the viscoelastic relaxation in styrene-based ionomers above 6 mol % of ions.

Dynamic mechanical analyses of ionomers derived from polystyrene-methacrylic acid were conducted by means of a torsion pendulum. Experiments were performed at 1-1.5°/min, under a positive nitrogen pressure. Loss tangent data are shown in Figure 12 for samples below the critical ion concentration. Two peaks were observed. One of these is due to the glass transition and increases with increasing ion concentration as expected. The upper peak, however, occurs at approximately the same temperature, although it may increase in intensity as the ion concentration increases. Tan δ for the lowest ion content (0.6) showed a continuous increase in tan δ as the flow region is approached. Samples above the critical ion concentration exhibit the same continuous increases in tan δ; however, no additional peaks are observed above room temperature.

This behavior is summarized in Figure 13. The lower curve is the glass transition temperature; the upper represents the second peak up to the critical ion concentration. Beyond that it represents (arbitrarily) the temperature at which tan δ = 0.5; this latter treatment is intended to illustrate the movement of corresponding behavior to higher temperatures as the ion concentration increases.

The small second peak at about 150°C for samples below 6 mol % ion content was attributed to "softening" of ionic multiplets which must persist to even higher temperatures based on viscoelastic relaxation data. Samples above 6 mol % did not show such a peak; this was described as being due to the fact that ionic clusters do soften as the temperature is increased, the transition is quite broad, and the clusters are viewed as changing structure continuously rather than over a narrow temperature range.

A final comment made about the process at 150°C is that it probably has a high activation energy since it did not shift position appreciably with changing frequency. Thus this transition may be comparable to the glass transition which is known to have a high (100-200 kcal/mol) activation energy.

Block Polymers

The previous section on regulation of copolymer sequence distribution covered two of three general classifications of polymers for this particular type of microstructure (i.e., perfectly alternating sequences and random distribution of sequences). The remaining type is the case where the sequence lengths of one or both comonomers achieve sufficient length so that glassy or crystalline aggregates occur in the polymer. This situation may arise on a microscale, due to low levels of block structure

arising from e.g. perturbations in the polymerization system, or on a
macroscale due to deliberate manipulation of synthesis conditions in
order to obtain polymers which have new sets of properties. This discussion will be primarily concerned with the applications of Thermal
Analysis to the latter (macroblock) type of polymer.

The impetus for much of the extensive activity in the field of block
copolymers appears to have come from the development of the styrene-
butadiene-styrene and related polymers. More recent activity in this
field has been concerned with a variety of additional systems including
polyurethanes,[20] polymers containing dimethyl siloxanes,[21] and
ethylene-propylene copolymers. For the present discussion the styrene-
butadiene system serves adequately to illustrate the many opportunities
for thermal analysis to contribute in the development of new polymer compositions. These materials possess properties which make them commercially
useful as thermoplastic elastomers.[19] (Figure14) This is made possible
by the fact that the styrene blocks form aggregates which act as physical
crosslinks over a temperature range, and the butadiene blocks convey
elastomeric properties to the system. These polymers are unique, however,
in that they can be made processable by heating the system above the glass
transition temperature of the styrene blocks. A model for the microstructure of such polymers is shown in Figure 15.[9] These materials
exhibit many of the properties of conventionally cross-linked elastomers
since the polystyrene aggregates act as both physical crosslinks and filler
particles. Further, entanglements involving the elastomeric blocks contribute additional physical cross-links to the system.

As might be anticipated from the previous comments, block polymer
systems offer the opportunity for highly useful application of a variety
of thermal analytical procedures. Two of these are illustrated in Figures
16 and 17.[10] The DSC data were obtained with a DuPont 900 Thermal Analyzer
(20-30 mg; compressed samples; polyamide reference material; samples re-
cycled 3-4 times to get reproducible data). DSC is an ideal technique for
determining transition behavior in these polymers since the Tg values for
the polystyrene segments and the polybutadiene segments are clearly shown
(Figure 16) and agree closely with independent viscoelastic analyses based
on the torsion pendulum (Figure 17). The two phase nature of these systems
is clearly evident and it is seen that the individual segment Tg's remain
constant over the entire copolymer composition range. This is in contrast
to typical random copolymers where one, composition-dependent Tg value is
observed.

The sharp, two phase nature of the styrene-butadiene system, plus
the demonstration that the molar heat capacity change (ΔCp) at Tg is the
same for the homopolymers, make it possible to determine copolymer composition from the DSC thermogram (i.e. since the magnitudes of the Tg
steps are proportional to the number of mole units of each component, the
ratio $\Delta B/\Delta S$ [see Figure 18] gives the copolymer composition).

The success of this method for these systems is shown in Table 3.
Reasonable agreement of DSC and NMR compositional analyses was also demonstrated for the styrene-isoprene and styrene-butadiene-isoprene systems.
In the latter case, a single Tg region was observed for the diene block
segments which was about midway between the Tg's of the two elastomeric
homopolymers. This was interpreted as an indication of relatively compatible phase blending of these two "soft" segments in the same polymer
chain, as contrasted with the separate phase behavior of block polymers
containing only a "hard" styrene phase and a single diene "soft" phase.

220

A. Analyses of Block Copolymers by Different Thermal Analytical Techniques

The previous discussion of block copolymers indicated that the major transition behavior in these polymers could be readily determined and interpreted. Upon reading the accounts of a number of such investigations, one might get the impression that the evaluation of transition behavior in these, and other polymer systems, is a straightforward matter,

TABLE 3

DSC RESULTS FOR STYRENE-BUTADIENE POLYMERS

Polymer Designation	Tg (B)* (°C)	Tg (S)* (°C)	Mole % S		
			DSC	NMR	Stoc.
S	--	111	100	100	100
S/B$_1$	-4	100	68	72	74
S/B$_2$	-1	104	55	55	59
S/B$_3$	-2	100	45	43	49
S/B4	2	103	24	36	39
S/B$_5$	0	99	18	19	19
B	-1	--	0	0	0
B/S	-2	105	82	78	80

*DSC values at 30°C/min. heating rate

and a relatively complete chapter in the state of knowledge concerning polymers. This is not completely the case, as is evident from a consideration of the complete microstructure of the block copolymer systems as well as a study of the literature regarding the determination of transition phenomena in polymers by thermal methods. With regard to the point about microstructure, for example, it is recognized that polybutadiene, which forms one of the segments in an SBS copolymer can potentially have three configurations (cis, trans and vinyl). Each of these has its own set of transition properties. The work cited earlier assigns only a single Tg value to the butadiene segments.

It has been shown, based upon the use of individual thermal techniques as well as the concerted use of several procedures, that the "transition" behavior of polymers and copolymers is often complex and sometimes subject to considerable differences of opinion regarding interpretation. It seems clear that some of the features which may be observed in "routine" thermal analyses of polymers are not readily assigned or explained.

A comprehensive investigation of SBS and SIS block copolymers conducted by Miller[11] illustrates the above point. The following material from that paper is presented here for the following purposes: (a) to illustrate the additional information which can be obtained by various techniques, (b) to indicate certain unusual observations present in the data and, (c) to stimulate discussion of this research area.

Figures 19, 20 and 21 show the thermograms obtained from the analysis of two common block copolymer systems by DTA, TDA (thermal depolarization analysis) and DMA (dynamic mechanical analysis). The least common of these techniques is TDA. A brief description is therefore in order because of the unusual characteristics and implications of the TDA thermograms for some polymer systems. The following description of the essential features of TDA is taken directly from Miller's publication.[11] "As the birefringence of a material changes with temperature, an increase in light level has been interpreted as that due to the recrystallization or more perfect ordering of areas within a polymer, whereas a decrease in the light level or birefringence is usually associated with the fusion or melting of an ordered area. The generation of peaks in the continuous plot of depolarized light transmission vs temperature is a frequent occurrence, and these peaks are due, for the most part, to increasing crystalline order during the small temperature range just prior to final fusion." It should be noted that the general assignment of peaks in the TDA curve to first order transitions should be approached with caution since other factors, e.g. form birefringence[23] could also be involved in some cases.

A comparison of the transition temperatures observed for SBS and SIS polymers by various thermal methods is shown in Table 4.[11] One of the striking features of this table is the multiplicity of transition regions for some of these polymers. The following conclusions from

TABLE 4

TRANSITION TEMPERATURES BY VARIOUS THERMAL METHODS [11]

Polymer	Transition Temperature (°C)				
	DTA	TDA	DMA*	Volume	Lit.
Poly(styrene-butadiene-styrene)				-148	
	-80	-95	-95	- 78	-80
		36	(-72)	34	
		65	89		
			121		
Poly(styrene-isoprene-styrene)	-57	-57	-57	- 59	-64
		- 6	(-40)	- 29	

*Temperatures in parentheses are damping maxima.

Miller's study are presented below as a summary of the work.

1. A clearer interpretation of block copolymer morphology is possible if the thermal response is studied by a variety of techniques.

2. DMA, DTA and DSC respond to more easily identified processes than do TDA or dilatometry.

3. Both volumetric and TDA techniques "clearly illustrate ordering or disordering processes that occur at temperatures above Tg and in this sense are more indicative of the morphological changes than are DSC and DMA."

4. The literature Tg assignment for the SIS polymer appears question-
able because of the TDA response near -57°C.

5. The onset temperature of the mechanical loss peak may be more indicative, in many cases, of the transitional change than is the
temperature of the loss peak since it correlates well with low-
frequency measurements.

The work described above, as well as other extensive studies by Miller,
represent an important contribution in the area of analysis of polymer transitions and microstructure.

Compatible Polymer Blends

Polymer blending is a common route to improving the properties of
polymer systems. Some examples of this are the modification of impact
properties in plastics via blending with elastomers, and the strengthing
of weak polymers with fibers. Most polymers are not truly miscible due
to basic thermodynamic considerations. However, a few pairs of polymers
have been found which are considered to be miscible on the basis of the
normal criteria used to assess compatibility.[12,16] Thermal analytical
procedures have proven quite useful for the analysis of such polymer blends,
since one of the most common tests of compatibility is the appearance of a
single Tg value intermediate between those of the components of the blend.
A recent example of this type of analysis is included here for illustrative
purposes.

Considerable effort has been expended on the analysis of compatible
polymer blends with regard to practical utilization of their properties
as well as achieving an improved understanding of the factors which are
responsible for the observed compatibility. MacKnight and co-workers
have conducted a series of investigations of one such blend, poly(2,6-
dimethyl-1,4-phenylene oxide) (PPO) and atactic polystyrene (PS).
References to those studies are contained in Reference (12) of this discussion. On the basis of DSC (single Tg intermediate between PPO and PS),
dynamic mechanical analysis (two partially merged relaxation peaks near
the DSC Tg) and dielectric relaxation (one peak corresponding to Tg, but
of different composition dependence than DSC or DMA), it was concluded
that PPO and PS are miscible but probably not on a segmental level.[12]
The data are shown in Figure 22.

Subsequent studies of the influence of polymer polarity on compatibility
established the fact that blends of PPO and poly(styrene-co-parachlorostyrene)
(PCS) could be compatibile or incompatible depending upon the chlorine content of the PCS. The application of DSC and dielectric relaxation to these
systems have very recently been reported.[12] As shown in Figure 23, the
polymers which contained the highest levels (mole percent) of parachlorostyrene [PCS-1 (100) and PCS-3 (68)] appear to be incompatible with PPO
since two Tg values were observed in the blends. However PCS-5, which contained the lowest level (47 mole %) of parachlorostyrene, was compatible
with PPO as shown by the single Tg intermediate between that of PPO and
PCS. An interesting observation relative to the compatible PPO-PCS 5 blends
is that even though the Tg of the system is proportional to PPO content, the
presence of PPO does not significantly effect the frequency and temperature
dependence of the dielectric loss peak which arises from dipolar reorientation
in PCS.

Problems Encountered During Thermal Analysis of New Materials

It is aboundantly clear that polymer systems may have a multiplicity of microstructures depending on the method of synthesis, the nature of the monomers and the catalyst employed. The previous discussion has been concerned with thermal analytical detection of various structural features associated with deliberate manipulation of polymerization conditions. Unanticipated structural and compositional variations can also be encountered as the result of a number of factors including the following: unknown aspects of new catalyst systems, or variations in the catalyst during polymerization; variations in conversion; reactor mixing problems, etc. This type of variability is sometimes directly suggested by particular characteristics of the thermograms. In other cases it may go unnoticed unless uncovered in other characterization studies (e.g. fractionation), or suggested by physical property variations or anomalies during e.g. routine polymer screening studies.

The final part of this presentation will illustrate this type of complication using as an example the results of Collins and co-workers for butadiene/acrylonitrile copolymers. Their investigation recently showed that two transitions were present in all commercial non-crosslinked B/AN copolymers of less than 35 percent acrylonitrile even though this fact had previously gone undetected.[13] This same type of relationship is illustrated for experimental polymers in Figures 24 and 25 which also show another example of the agreement among various thermal analytical procedures. The dependence of transition behavior on conversion was carefully analyzed. In every case where there was a second Tg, it moved to lower values as conversion proceeded.

It was presumed, based upon these observations, that the two transitions in samples containing less than 34 percent AN were due to two distinct phases in the polymer. A simple fractionation procedure was employed to examine this possibility. It was found that the two phases in such a polymer were separable by solution fractionation using carbon tetrachloride as a solvent. DTA scans (1°/min; DuPont 900DTA) were obtained on the soluble and insoluble fractions and are compared to the original polymer (21% AN) in Figure 26. Although this one step fractionation was not completely effective, separation of the two species can be recognized via comparison of the thermograms. Rheovibron data for the same fractions verified that the two transitions in the original sample can be physically separated. Tg variation was used to show that, in NBR/PVC blends, PVC acts like a selective solvent for only the NBR component which has the highest acrylonitrile content.

Collins speculated that if incompatibility was the basis for the multiple transitions observed in NBR, a physical mixture of two NBR samples which differed significantly in Tg should show both Tg values. A test of this point was made by Ambler.[14] The results (Figure 27) verify the prediction of Collins.

Ambler also conducted several polymerizations in which varied incremental additions of monomer were utilized. He observed that the breadth of the Tg interval varied significantly depending on the manner of monomer addition. Based on this observation, he suggests that the characteristic transition curve shape appears to be an indicator of the distribution of copolymer species in the polymer. The observation regarding the relationship between the shape of the Tg region and monomer sequence distribution is similar to relationships previously proposed for ethylene propylene copolymer systems.[22]

SUMMARY

The examples presented in this discussion indicate the wide acceptance and utility of thermal analytical techniques in both the research and applied areas of polymer science. A significant trend in the applied area is toward more complex polymer compositions, including blends and composites. The multicomponent nature of these systems will in many instances require the application of several characterization procedures to elucidate the composition, morphology and structure-property relationships. It seems clear that several of the basic thermal analytical techniques will play a continuing and important role in these investigations.

ACKNOWLEDGMENT

Several prominent investigators in the field of Thermal Analysis were kind enough to assist the author via copies of their published work, current research summaries, etc. Grateful acknowledgment of such assistance is extended herewith to: Professors J. F. Johnson (University of Connecticut) and W. J. MacKnight (University of Massachusetts), and Drs. E. M. Barrall II (IBM Corp.), J. Chiu (DuPont Co.), G. W. Miller (Owens-Illinois Corp.), F. Noel (Imperial Oil, Ltd.) and R. A. Pethrick (University of Strathclyde).

Thanks are due also to the American Chemical Society; John Wiley and Sons, Plenum Publishing Co., Thermochimica Acta and Marcel Dekker Co. for use of material appearing in the publications cited herein.

REFERENCES

1. C. E. Wilkes, M. J. Pelko and R. J. Minchak, J. Polymer Sci., Symposium 43, 97 (1973).
2. N. W. Johnston, Polymer Preprints 14, 634 (1973).
3. M. Hirooka and T. Kato, J. Polymer Sci., (Letters) 12, 31 (1974).
4. A. Eisenberg and M. Navratil, Macromolecules 6, 606 (1973).
5. M. W. Williams, Polymer Preprints, 14, 896 (1973).
6. S. L. Cooper, C. L. Marx and D. F. Caulfield, Macromolecules 6, 344 (1973).
7. S. L. Cooper, C. L. Marx and D. F. Caulfield, Polymer Preprints 14, 890 (1973).
8. A. Eisenberg and M. Navratil, Macromolecules 7, 90 (1974).
9. E. T. Bishop and S. Davison, Polymer Symposia 26, 59 (1969).
10. R. M. Ikeda, M. L. Wallach and R. J. Angelo in Block Polymers, S. Aggarwall, Ed., p. 43, 1969, Plenum Press, New York.
11. G. W. Miller, Thermochimica Acta 4, 425 (1972).
12. F. E. Karasz, W. J. MacKnight and J. J. Thacik, Polymer Preprints, 15, 415 (1974).
13. E. A. Collins, A. H. Jorgensen, and L. A. Chandler, Rubber Chem, and Tech. 46, 1087 (1973).
14. M. R. Ambler, J. Polymer Sci. (Chemistry) 11, 1505 (1973).
15. S. L. Cooper and R. W. Seymour, Macromolecules 6, 48 (1973).
16. D. R. Paul and J. O. Altimarano, Polymer Preprints 15, 409 (1974).

17. W. J. MacKnight, Polymer Preprints 14, 813 (1973).
18. A. Eisenberg, Macromolecules 4, 125 (1971).
19. G. Holden, et. al., Polymer Symposia 26, 37 (1969).
20. S. L. Cooper and R. W. Seymour, Macromolecules 6, 48 (1973)
21. M. Matzner, et. al., Applied Polymer Symposia, No. 22, 143 (1973).
22. J. J. Maurer, Rubber Chem. and Tech. 38, 979 (1965); Eastern Analytical Symposium, New York, 1967; University of Utah Polymer Conference Series, 1970.
23. E. M. Barrall, II, private communication.

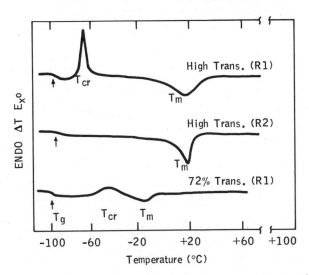

Fig.1 DSC analysis of selected polypentenamers[1]

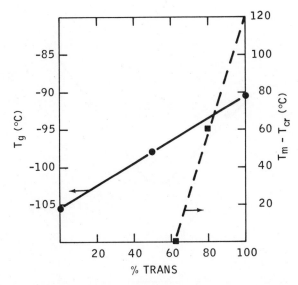

Fig.2 Transitions vs polypentenamer trans content[1]

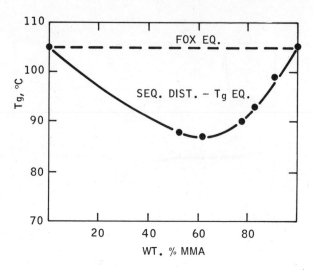

Fig.3 Tg evaluation of AN/MMA copolymers[2]

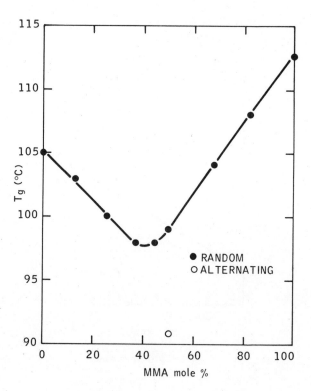

Fig.4 Tg vs composition: styrene/MMA copolymers[3]

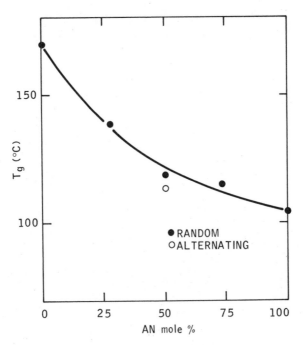

Fig.5　Tg vs composition: ∝-methyl styrene/AN copolymers[3]

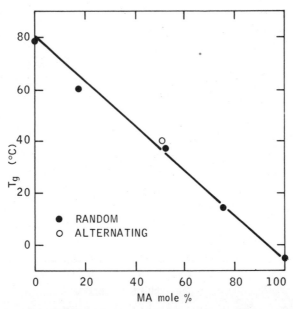

Fig.6　Tg vs composition: vinyl chloride/methyl acrylate copolymers[3]

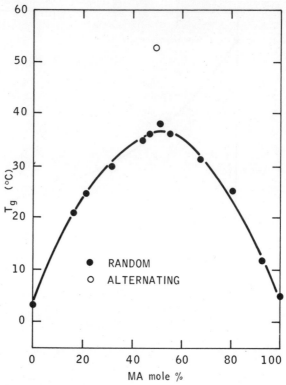

Fig.7 T_g vs composition: vinylidene Cl/methyl acrylate copolymers[3]

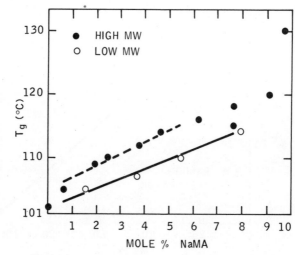

Fig.8 T_g vs ion content: styrene/MA copolymers[4]

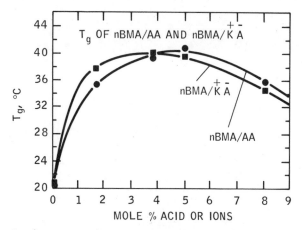

Fig.9 Tg vs acid or ion content: styrene/n BMA[5])

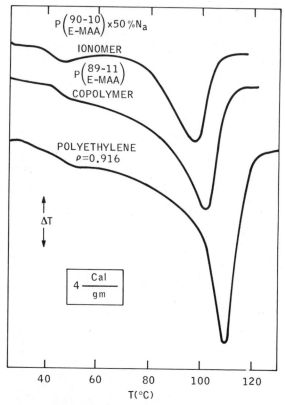

Fig.10 Effect of copolymerization and ionization on PE/MAA crystallinity[6])

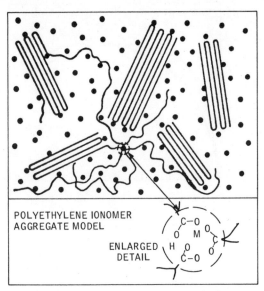

Fig.11 Aggregate model of ionomer morphology: schematic
representation[7]

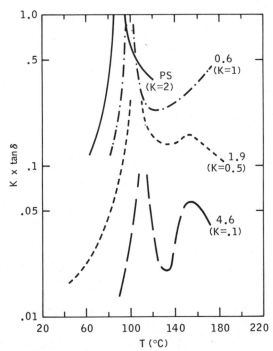

Fig.12 Transitions in styrene/MA copolymers below critical
ion concentration[8]

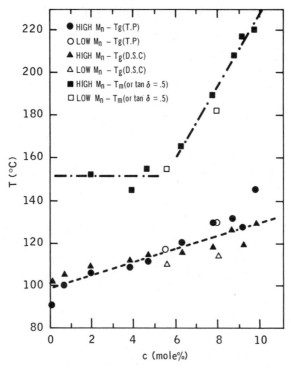

Fig.13 Effect of ion concentration on styrene/MA transitions[8]

	THERMOSETTING	THERMOPLASTIC
RIGID	EPOXY PHENOL-FORMALDEHYDE UREA-FORMALDEHYDE HARD RUBBER	POLYSTYRENE POLYVINYL CHLORIDE POLYPROPYLENE
FLEXIBLE	HIGHLY LOADED AND/OR HIGHLY VULCANIZED RUBBERS	POLYETHYLENE ETHYLENE-VINYL ACETATE COPOLYMER PLASTICIZED PVC
RUBBERY	VULCANIZED RUBBERS (NR, SBR, IR, ETC.)	THERMOPLASTIC ELASTOMERS

Fig.14 Classification of polymers[9]

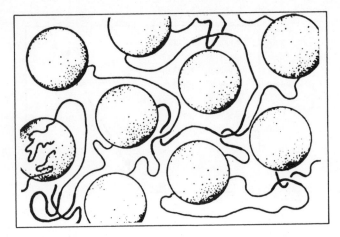

Fig.15 Model of ABA block polymer structure (~25% END block)[19]

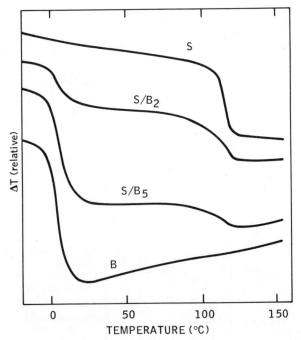

Fig.16 Typical DSC curves for S/B/S block polymers[10]

234

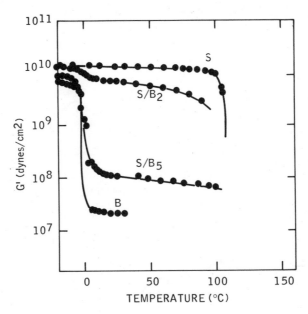

Fig.17 Torsion pendulum shear modulus for SBS block copolymers[10]

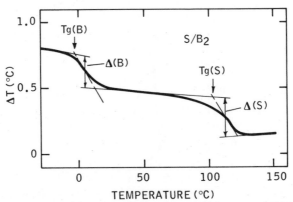

Fig.18 Method of analyzing DSC thermograms[10]

235

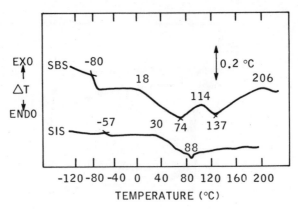

Fig.19 DTA characteristics of SBS and SIS copolymers[11]

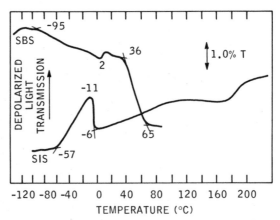

Fig.20 Thermal depolarization analysis: SBS and SIS polymers[11]

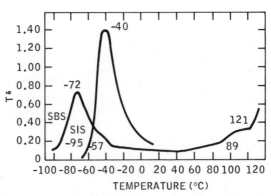

Fig.21 dynamic mechanical analysis: SBS and SIS copolymers[11]

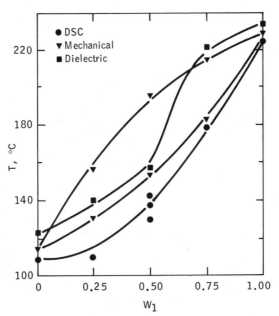

Fig.22 Tg by different techniques: PPO-PS mixtures[12]

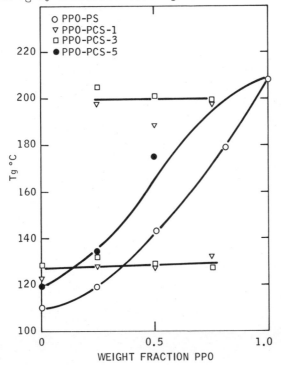

Fig.23 DSC Tg : PPO-PS mixtures and PPO-PCS mixtures[12]

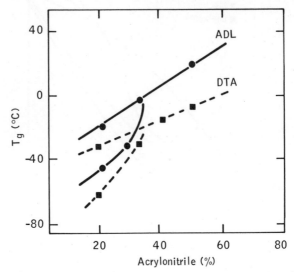

Fig.24 Tg vs copmosition: B/AN copolymers[13]

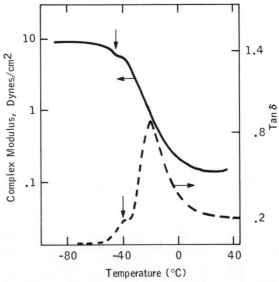

Fig.25 Dynamic properties of 22% AN copolymer[13]

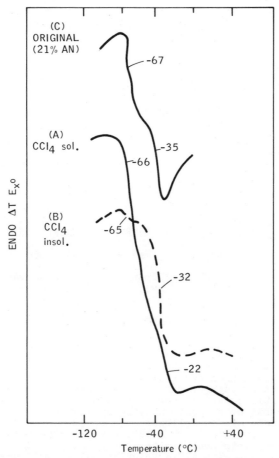

Fig.26 DTA of B/AN copolymer fractions[13]

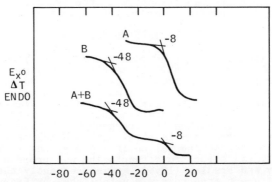

Fig.27 Illustration of B/AN blend incompatibility[14]

239

THERMAL ANALYSIS OF SEPIOLITE AND TREATED EQUIVALENTS

R. Otsuka, N. Imai and T. Sakamoto

Department of Mineral Industry, Waseda University,
Tokyo, Japan

ABSTRACT

Dehydration and structural changes on heating of natural
sepiolite were studied by thermal and other techniques. A good correlation was found between the water experimentally determined at
different temperatures and that inferred for the ideal formula based
on the Brauner-Preisinger model rather than the Nagy-Bradley model.
In addition, natural sepiolite was treated in various ways,
and the products were examined by X-ray and thermal techniques. The
results obtained are as follows:
1) After treatment with NaOH solution at room temperature, sepiolite
was transformed into loughlinite.
2) After treatment with $AlCl_3$ solution below 250°C under hydrothermal
conditions, sepiolite was changed into a well-crystallized kaolinite.
3) Under water pressures to 1000 Kg/cm^2 and temperatures ranging ·from
200-700°C, sepiolite was transformed into hydrated talc, and then
talc. P-T relationships observed for sepiolite were also obtained.

1. INTRODUCTION

Sepiolites are light hydrous magnesium silicates with an amphi-
bole chain structure, classified as clay minerals. Sepiolite may
occur either in large fibrous crystals associated with hydrothermal
deposits, or in cryptocrystalline masses of a sedimentary origin.
The crystalline fibrous variety is called para or α-sepiolite, while
the amorphous and cryptocrystalline compact variety is called β-
sepiolite. It also occurs as supergene alteration products of ser-
pentinite in its cracks. In particular, sepiolite has been recently
discovered in the continental shelf and deep-sea sediments.
The crystal structure of sepiolite was proposed for the first
time by Nagy and Bradley (1955), and subsequently the similar but
slightly different structure was established by Brauner and
Preisinger (1956). The unit cell dimensions and space group according to them are listed in Table 1.

Table 1   Unit cell dimensions and space group of sepiolite

	(1)	(2)
a sin β	13.4	-
a	-	13.4
b	27.0	26.8
c	5.3	5.28
S.G.	C2/m	PnCn

(1) Nagy and Bradley (1955), (2) Brauner and Preisinger (1956)

The half-unit-cell contents for the Nagy-Bradley model are $(Si_{12})(Mg_9)O_{30}(OH)_6(OH_2)_46H_2O$ and $(Si_{12})(Mg_8)O_{30}(OH)_4(OH_2)_48H_2O$ for the Brauner-Preisinger model, where OH_2 denotes bound water and H_2O represents water held in channels. Recently, certain improvements have been made to the structure proposed by Brauner and Preisinger (Raututreau et al., 1972).

It is well known that water in sepiolite consists of the following four groups: (1) hygroscopic water, (2) zeolitic water in the channel of structure, (3) water molecules bound on the edge of octahedral sheet-bound water-, and (4) hydroxyl groups associated with octahedral sheet (Bradley, 1940; Nagy and Bradley, 1955; Martin Vivaldi and Cano Ruiz, 1956a, b; Preisinger, 1957 and 1963). However, regarding thermal behaviours of sepiolite in relation to the bonding character of water various opinions have been expressed. On the other hand, sepiolite has a wide range of industrial applications because of its large specific surface (ca. 150 m^2/g) and the cavity channel in its structure (Robertson, 1957; Preisinger, 1963). Recently, it has been used as a support material in catalysis and in gas chromatography; it may prove useful in the treatment of exhaust gases (Preisinger, 1963). According to Müller and Koltermann (1965), different adsorption properties of sepiolite show that the channel in the sepiolite structure are 6-8A in diameter, and that sepiolite has the properties of a molecular sieve. Therefore, it may be considered that sepiolite is an interesting mineral from the view points of both mineralogy and industrial application.

Sepiolite is a rare mineral in Japan. However, several occurrences have lately been reported (Muraoka et al., 1958; Muchi et al., 1965; Minato, 1966; Imai et al., 1966 and 1967). An extensive mineralogical study of Japanese sepiolites is being carried out in our laboratory.
In the present paper, simultaneous DTA-TG-DTG curves for natural sepiolites and treated equivalents in variuos ways are given. The interpretation of these curves have been supported by information obtained from other techniques, such as chemical, X-ray, infrared and electron microscopy methods.

2. NATURAL SEPIOLITE

Simutaneous DTA-TG-DTG curves for sepiolites from Kuzu, Tochigi Pref., Japan, and from Eskisehir, Turkey are given in Fig. 1. X-ray diffraction patterns for both specimens show extreme cases in crystallinity as shown in Fig. 2. Their chemical compositions and structural formulae are given as follows: Kuzu specimen, SiO_2 52.85, Al_2O_3 1.03, TiO_2 tr., Fe_2O_3 0.04, FeO 0.01, MnO 0.01, CaO 0.51, MgO 23.74, $H_2O(+)$ 9.04, $H_2O(-)$ 12.67, Total 99.89 and $(Si_{11.79}Al_{0.21})_{12}(Al_{0.06}Fe^{3+}_{0.01}Mg_{7.89})_{7.96}O_{32}Ca_{0.12}$; Eskisehir specimen, SiO_2 52.33, Al_2O_3 0.65, Fe_2O_3+ FeO 0.15, CaO 0.20, MgO 23.88, Na_2O 0.05, K_2O tr., $H_2O(+)$ 9.28, $H_2O(-)$ 13.01, Total 99.55 and $(Si_{11.86}Al_{0.14})_{12}(Al_{0.03}Fe_{0.03}Mg_{7.91})_{7.97}O_{32}Ca_{0.05}Na_{0.02}$. The formulae were calculated assuming that the dehydrated half-unit cell contains 32 oxygens according to the Brauner-Preisinger model.

On the basis of examinations by many investigators including the present writer and his co-workers (Caillère and Hénin, 1957, 1961; Hayashi et al., 1969; Imai et al., 1969; Kulbicki and Grim, 1959; Martin Vivaldi and Cano Ruiz, 1956; Martin Vivaldi and Fenoll Hach-Ali, 1970; Preisinger, 1957, 1963; Otsuka et al., 1966, 1970); these curves may be interpreted as below.

In the lower temperature region below 200 °C a large endothermic reaction is observed. The results of continuous high-temperature X-ray diffraction and IR analysis indicate that the structure is not modified by loss of water below 200°C (Imai et al., 1969, Hayashi et al., 1969). This water is easily regained. Therefore, the first endothermic peak and corresponding weight loss is attributed to elimination of hygroscopic and zeolitic water. The amount of water loss varies between 8-15%, depending on relative humidity (Martin Vivaldi et al., 1959).

A well-crystallized sepiolite shows usually two endothermic effects at about 300°C and 500°C in the central temperature region (Curve (2), Fig. 1). The TG-DTG curves represent a two-step weight loss of 5.4% over this range, in good agreement with the theoretical value of bound water. Owing to the dehydration of bound water above 250°C, the structure gradually changes into sepiolite anhydride, $Mg_8Si_{12}O_{30}(OH)_4$ (Preisinger, 1963). The X-ray and IR data obtained for heated sepiolite suggest that sepiolite and sepiolite anhydride coexist in this temperature region. Anhydride is stable up to 750°C, at which temperature the final endothermic reaction commences. According to the Martin Vivaldi and Cano Ruiz (1956b), the occurrence of two peaks on the DTA curve and two plateaux on the TG curve in this region has been attributed to a change in the bonding energy of bound water because of the structural alteration at 350°C. However, further examination will be needed in order to clarify the mechanism of

a two-step dehydration of bound water. In the case of a poorly
crystallized sepiolite (Curve (1), Fig. 1), the elimination of bound
water proceeds rather continuously.
      The final endothermic effect at about 800°C is accompanied by
the weight loss of 2.7%, which agrees well with the content of
hydroxyls indicated by the structure of Brauner and Preisinger
(956). This endothermic peak is followed immediately by the strong
sharp exothermic peak at about 820-830°C which is due to recrystallization of clinoenstatite, $MgSiO_3$. The c-axis of clinoenstatite has the same length and direction as the c-axis of sepiolite
(Preisinger, 1963).
      In Table 2 the loss of water found experimentally at different
temperature is compared with that expected from the two ideal formulae
based upon either the Nagy-Bradley model or the Brauner-Preisinger
model. The results appear to support the latter model.

Table 2   Observed water losses (%) of sepiolite
compared with those expected from the ideal formulae

	(1)	(2)	(3)	(4)
Zeolitic water	11.0	13.0	8.1	11.1
Bound water	5.4 2.7 2.7	5.5	5.4	5.5
Hydroxyls	2.7	2.5	4.1	2.7

(1) Kuzu, Tochigi Pref. (Imai et al., 1966)
(2) Eskisehir, Turkey (Otsuka u. a., 1973)
(3) Ideal sepiolite (Nagy and Bradley, 1955)
(4) Ideal sepiolite (Brauner and Preisinger, 1956)

3. SEPIOLITE TREATED IN VARIOUS WAYS

      Natural sepiolite was treated in following various ways. The
products were examined by means of thermal and other techniques.

3.1   Treatment with NaOH solution at low temperature

      Sepiolite from Kuzu was treated with 6N NaOH solution for one
hour at 92±5°C in a water bath and for 24 hrs. at room temperature
(Imai et al., 1969). The products, denoted by L_1 and L_2, were examined by X-ray and thermal methods.
X-ray diffraction: As shown in Fig. 3, X-ray diffraction patterns
for L_1 and L_2 are not identical with that of original sepiolite (S_0)

namely, the 12.2A reflection of sepiolite was replaced by the 12.9A reflection, and the reflections at 7.6A (130), 6.5A (220), 4.46-4.50 A (060), 3.88A (260), 3.08-3.09A (171) and 2.63A (281) which are characteristic to loughlinite were recognized. Also, the measured powder data have no distinct departure from those of natural loughlinite from Sweetwater County, Wyoming (L_0).

DTA, TG and DTG: Simultaneous DTA-TG-DTG curves for L_0 and L_1 are shown in Figs. 4 and 5. As seen from these figures, the curves for both L_0 and L_1 are quite different from that of S_0, although Fahey et al.(1960) stated that the DTA curve for natural loughlinite is similar to that of sepiolite. The TG-DTG curves for L_1 reveal that the weight loss is gradual over a wide temperature range from 80 to about 710°C, indicating that dehydration proceeds gradually over this range, provided that weight loss depends only upon dehydration. However, an attentive examination seems to reveal three steps of weight loss. The first step of weight loss is abrupt, starting at 80°C and terminating at about 250°C, as represented by the steep slope of the TG curve, corresponding to the broad DTG doublet having maxima at 120°C and 180°C. The rate of weight loss in the second step is low from 250 to 620°C. The third step of weight loss occurs in a narrow temperature interval from 620 to 720°C. The DTA curve shows a strong endothermic doublet having maxima at 100 and 180°C, and small endothermic effects at 680 and 700°C, followed by a very sharp and strong exothermic peak at 760°C. It seems that the marked endothermic peak at 400°C is attributed to dehydration of brucite contained in the product. DTA-TG-DTG curves for L_0 (Fig. 5) are similar to those for L_1 (Fig. 4), although not strictly identical.

In general, the peak temperatures on the DTG and DTA curves for L_0 shift towards lower temperature side as compared with those for L_1. A marked exothermic effect of about 320°C on the DTA curve for L_0 may be caused by the combustion of organic matter of shale contained as impurity in the sample examined.

Thus, the present experimental work clearly indicates that sepiolite can be trasformed into loughlinite by treatment with NaOH solution at temperatures lower than 100°C, even at room temperature, under appropriate chemical conditions. In addition, the resultant loughlinite could be reversibly returned to sepiolite by treatment with cold 1N $MgCl_2H_2O$ solution as was already confirmed in the case of natural loughlinite by Fahey et al. (1960). These facts strongly suggest that the reaction involves cation exchange of $2Mg^{++}$ by $4Na^+$, and it may be represented by the following chemical equation:

Sepiolite
$$(H_2O)_8Mg_8(OH_2)_4(OH)_4Si_{12}O_{30}(c) + 4Na^+(aq) + 2H_2O + 2OH^-(aq) \longrightarrow$$

Loughlinite   Brucite
$$(H_2O)_{10}Na_2(Na_2Mg_6)(OH_2)_4(OH)_4Si_{12}O_{30}(c) + 2Mg(OH)_2(c)$$

where (c) and (aq) denote crystalline phase and ion in aqueous solution, respectively.

3.2 Hydrothermal treatment with $AlCl_3$ solution

Sepiolite from Kuzu was hydrothermally treated with $1N$ $AlCl_3 H_2O$ solution below 250°C, using a reaction vessel of Morey type. In this type of equipment, the pressure of the system is always governed by the saturated vapor of water. Treated sepiolites were examined by X-ray, thermal and electron optical methods. Dissolved Mg in the solution was determined by atomic absorption spectrometer.

In Fig. 6 and Fig. 7 are given X-ray diffraction patterns and DTA-TG curves for sepiolite from Kuzu after treatment for 2 hrs, 3 hrs, 6 hrs, 12 hrs, 1 day, 3 days, 5 days, 7 days and 10 days at 250°C, where the water vapor pressure of the system is estimated at approximately 40 Kg/cm^2.

X-ray diffraction: The X-ray patterns show a remarkable change even after 2 hrs' treatment. All reflections for the untreated sepiolite disappear or considerably weaken, while new reflections at about 4.4A and 1.5A appear. After 6 hrs' treatment a broad peak at 7.2A occur. After 1 day's treatment, the X-ray patterns for the product are very similar to those of kaolin mineral. With the increase of treatment time, the reflections become gradually sharper and well-resolved. After 5 days' treatment the patterns of the product correspond to a well-crystallized kaolinite, $Al_2Si_2O_5(OH)_4$.

DTA and TG: The DTA and TG give very similar results to those obtained by the X-ray method. The DTA-TG curves for the products are different entirely from those of the untreated sepiolite even after 2 hrs' treatment. A broad endothermic peak at 500-600°C and a small exothermic peak at about 980°C appear on the DTA curve. The TG curve exhibits a continuous weight loss in the wide temperature range from room temp. to 1,000°C. With the increase of treatment time, the DTA peaks become larger and sharper, except for the endothermic effect below 100°C, ascribed to loss of physically held water. The corresponding weight loss between 400-600°C occurs more rapidly. After 5 days' treatment the DTA-TG curves for the product correspond well to those of a well-crystallized kaolinite.

These results are supported by information obtained by the electron optical observation, and chemical analysis of dissolved Mg in the solution.

Therefore, it may be concluded that sepiolite is changed into kaolinite after treatment with $AlCl_3$ solution under hydrothermal conditions.

3.3 Hydrothermal treatment at elevated pressure

In order to investigate phase transformations of sepiolite under hydrothermal conditions, sepiolite from Kuzu was treated in the temperature range from 150 to 700°C under water pressures of 250, 500 and 1,000 Kg/cm², respectively (Otsuka et al., 1974). In Fig. 8 and Fig. 9 are given X-ray diffraction patterns and DTA-TG curves for the products obtained by the treatment at P_{H_2O} 500 Kg/cm², for 5 days.

X-ray diffraction: X-ray patterns for the product remain unchanged after treatment at 325°C. With increasing temperature, the reflection of 9.7A, appearing at 350°C, becomes more distinct and sharper together with other reflections at 4.73A and 3.16A. On the other hand, the reflections of the untreated sepiolite gradually decrease in intensity, and they disappear at 450°C. The X-ray patterns for the product at 450°C are very similar to those of talc. However, the 002 reflection, corresponding to a spacing 9.6A, shifts towards low-2θ side from that of ordinary talc, usually appearing at 9.4-9.3A. The basal reflections at 9.6A (002), 4.73A (004) and 3.15A (006) do not form a simple rational series. The spacing of the 002 reflection remains unchanged when treated with ethylene glycol. Therefore, it is suggested that the product at 450°C corresponds to hydrated talc rather than ordinary talc. Mineralogical properties of naturally occurring hydrated talc have been lately reported by Brindley and Hang (1972) and Imai et al. (1973). Above 500°C, the intensities of the basal reflections increase remarkably, and the 002 reflection appears at 9.4-9.3A. Accordingly, it seems that the products above 500°C correspond to ordinary talc.

DTA and TG: The results of the DTA-TG are in good agreement with those of the X-ray examination. The DTA-TG curves for the product remain unchanged at 325°C. A new endothermic effect appears at 860°C on the DTA curve for the product at 350°C. This effect may be ascribed to dehydroxylation of hydrated talc. The curve clearly indicate that sepiolite and hydrated talc coexist in the temperature range from 350 to 400°C, and that the amount of sepiolite diminishes gradually with increasing temperature. The endothermic effect at about 860°C, accompanied by weight loss, may be caused by dehydroxylation of talc.

Therefore, under water pressure of 500 Kg/cm², sepiolite is stable below 330°C. Above this temperature, it transforms gradually into hydrated talc (330-460°C) and talc (460°C). P-T relationships observed for sepiolite are given in Fig. 10. This figure shows that an intermediate phase such as hydrated talc comes between sepiolite and talc in their stability field. Similar results were already obtained by Frank-Kamenetskiy et al.(1969), and in the case of stevensite by Otsuka et al. (1972).

4. CONCLUSIONS

In this investigation, regarding structural changes of sepiolite occurring during heating the similar results are those by other investigators were obtained. However, many problems remain unresolved, relating to the behaviour of the two-step dehydration of bound water.

Furthermore, the present investigation revealed that sepiolite can be easily transformed into various mineral species after treatment in various ways, even at room temperature. Progressive phase changes of treated sepiolite with increasing temperature or duration were recognized well by the combination of X-ray and thermal techniques. The DTA and TG gave generally very similar results to those by the X-ray diffraction method.

REFERENCES

Bradley, W. F. (1940), Am. Miner., 25, 405.
Brauner, K. and Preisinger, A. (1956), Mineralog. petrogr. Mitt., 6, 120.
Brindley, G. W. and Hang, P. T. (1972), Clays and Clay Minerals, 21, 27.
Caillere, S and Henin, S. (1957), In "Differential Thermal Investigation of Clays" (R. C. MacKenzie, ed.), Min. Soc., London, 231.
----------- and -------- (1961), In "The X-ray Identification and Crystal Structures of Clay Minerals" (G. Brown, ed.), Min. Soc., London, 325.
Fahey, J. J., Ross, M. and Axelrod, J. M. (1960), Am. Miner., 45, 270.
Hayashi, H., Otsuka, R. and Imai, N. (1969), Am. Miner.,53, 1613.
Imai, N., Otsuka, R., Nakamura, T. and Inoue, H. (1966), J. Clay Sci. Soc. Japan, 6, 30 (in Japanese).
--------, --------- and Nakamura, T. (1967), J. Japan. Assoc. Min. Pet. Econ. Geol., 57, 39.
--------, ---------, Kashide, H. and Hayashi, H. (1969), Proc. Inter. Clay Conf. Tokyo, 99.
--------, ---------, Nakamura, T. and Koga, M. (1969), Bull. Sci. Eng. Res. Lab., Waseda Univ., No. 46, 11.
-------- ---------, --------------, Tsunashima, A and Sakamoto, T. (1973), Clay Sci., 4, 175.
Kulbicki, G. and Grim, R. E. (1959), Mineralog. Mag., 32, 53.
Martin Vivladi, J. L. and Cano Ruiz, J. (1956a), Clays and Clay Minerals, 4, 173.
-------------------- and -------------- (1956b),   ibid., 4, 177.
-------------------- and Fenoll Hach-Ali, P. (1970), In " Differential Thermal Analysis" Vol. 1 (R. C. MacKenzie, ed.), Academic Press, London and New York, 553.
Minato, H. (1966), J. Clay Sci. Soc. Japan, 6, 22.
Muchi, M., Hoshino, Y. and Furusato, I (1965), J. Japan Assoc. Min. Pet. Econ. Geol., 53, 39.
Muller, K. P. and Koltermann, M. (1965), Z. f. anorg. allge. Chemie, 341, 36.
Muraoka, H., Minato, H, Takano, Y and Okamoto, Y. (1958), J. Miner. Soc. Japan, 3, 381.
Nagy, B and Bradley, W. F. (1955), Am. Miner., 40, 885.
Otsuka, R., Imai, N. and Nishikawa, M. (1966), Kogyo Kagaku Zasshi

_69_, 1967.
------------, Hayashi, H. and Imai, N. (1970), Bull. Sci. Eng. Res.
Lab. Waseda Univ., No. 47, 56.
------------, Imai, N. and Sakamoto, T. (1972), Mem. Sch. Sci. Eng.
Waseda Univ., No. 36, 37.
------------, Sakamoto, T. and Hara, Y. (1974). J. Clay Sci. Soc.
Japan, _14_, 8. (in Japanese).
Preisinger, A. (1957), Clays and Clay Minerals, _6_, 61.
--------------. (1963),   ibid.   , _10_, 365.
Rautureau, M., Tchoubar, C. and Mering, J. (1972), Proc. Inter. Clay
Conf., Madrid, 153.
Robertson, R. H. S. (1957), Chem. Ind., 1492.

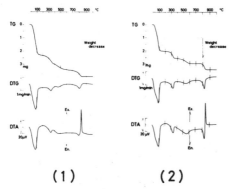

(1)   **(2)**

Fig. 1 Simultaneous DTA-TG-DTG curves
for sepiolite from: (1) Eskisehir,
Turkey, (2) Kuzu, Tochigi Pref., Japan

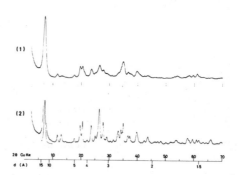

Fig. 2 X-ray diffraction patterns for
sepiolite from: (1) Eskisehir, Turkey,
(2) Kuzu, Tochigi Pref., Japan

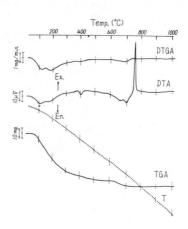

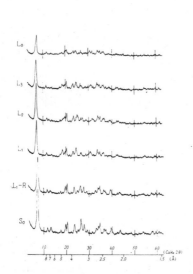

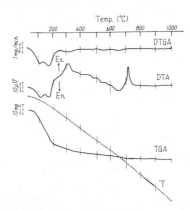

Fig. 3 X-ray diffraction patterns for sepiolite (S_0) and the treated equivalents. L_0: Natural loughlinite from Sweetwater County, Wyoming. L_1, L_2 and L_3: Treated equivalents of S_0 with 6N NaOH solution. L_1-R: Retreated equivalents of L_1 with 1N $MgCl_2H_2O$ solution.

Fig. 4 Simultaneous DTA-TG-DTG curves for L_1

Fig. 5 Simultaneous DTA-TG-DTG curves for L_0 (natural loughlinite from Wyoming)

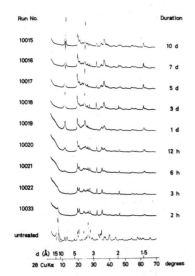

Fig. 6 X-ray diffraction patterns
for sepiolite and equivalents
treated with AlCl₃ solution under
hydrothermal conditions (250°C, 2
hrs., 10 days)

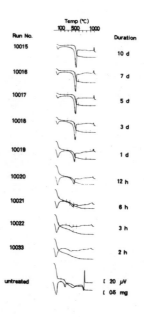

Fig. 7 Simultaneous DTA-TG
curves for sepiolite and its
equivalents treated with
AlCl₃ solution under hydrothermal conditions (250°C,
2 hrs., 10 days)

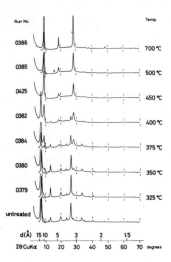

Fig. 8 X-ray diffraction patterns for sepiolite from Kuzu after hydrothermal treatment (P_{H_2O} 1,000 Kg/cm^2, 325-700 °C, 5 days)

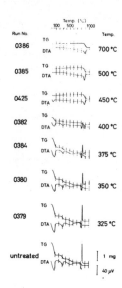

Fig. 9 Simultaneous DTA-TG curves for sepiolite from Kuzu after hydrothermal treatment (P_{H_2O} 1,000 Kg/cm^2, 325-700 °C, 5 days)

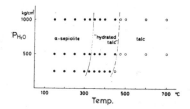

Fig. 10 P-T relationships observed for sepiolite

MICROCALORIMETRY APPLIED TO BIOCHEMICAL PROCESSES

E. J. Prosen, R. N. Goldberg, B. R. Staples, R. N. Boyd,
and G. T. Armstrong

Thermochemistry Section, Physical Chemistry Division,
Institute for Materials Research
National Bureau of Standards
Washington, DC, U.S.A.

ABSTRACT

A single-reaction-vessel, batch-type, conduction micro-calorimeter is described, having a volume of about 0.3 ml. The sensitivity in terms of electrical output voltage per unit of heat transfer power is 60 mV·W^{-1}, the lower limit of detectable thermal power is about 0.1 μW, and the lower limit of total heat that can be measured is <0.2 mJ. The calori-meter is calibrated electrically and by known chemical reactions such as the neutralization of HCl by NaOH in solution. The calorimeter has been applied to the measurement of enzyme catalyzed biochemical processes. On the basis of measurements made in aqueous buffer solutions, the hexokinase catalyzed phosphorylation of glucose has been used to assay glucose in the complex media human serum and blood plasma, with results that correlate well with customary clinical laboratory procedures. The growths of Enterobacter cloacae, Proteus rettgeri, and Klebsiella pneumoniae in the calori-meter under controlled conditions show characteristic energy evolution patterns. In a somewhat larger reaction cell, the fertilization and growth of Arbacia punctulata eggs were observed.

INTRODUCTION

The use of clinical laboratory tests of physiological fluids for specific components has been increasing rapidly in the past few years. There has been an increase in the number of tests performed for well-known particular sub-stances for which acceptable tests procedures are available. Tests for increasing numbers of substances are also needed by the physician and by the research biochemist. Because of this increasing demand for tests, the development of new techniques applicable to problems for which satisfactory solutions have not been found is of considerable interest in clinical and biological chemistry.

The universal occurrence of energy changes in chemical and physiological processes has caused calorimetry to be

253

considered as a potential procedure for clinical chemical analysis and for observing biological processes for some years. These applications of calorimetry require the use of a microcalorimeter, not only because most of the materials of interest are available only in very small quantities, but also because these quantities are frequently in an aqueous medium in very low concentrations. It is under these aqueous conditions that we wish to study many of the materials rather than removing them from the medium and studying their properties in the pure solid or liquid state or in concentrated solution. The investigator many have to work with micromoles or nanomoles of substance and with proportions of the order of 1 mg in 10^5 mg of water or other aqueous media. For clinical studies the quantities available are always small and the smaller the quantity required for each test, the larger the number of tests that can be made on the available quantity. The heats or enthalpies of biochemical reactions per mole of substance are not unusually small, but since the amounts or concentrations or both are small, the measured heats will be very small.

In biochemical research and in clinical studies of disease, it is necessary to perform measurements under various conditions and on many different samples before conclusions can be reached. In addition, the person carrying out such measurements will probably not have training specifically in calorimetry. These factors impose additional constraints on the characteristics of calorimeters suitable to this field of study.

In addition to great sensitivity and small sample capacity a microcalorimetry useful for routine biochemical and clinical studies should be simple and reliable in operation, rapid for single and sequential tests, sturdy and simple in construction, and inexpensive to produce.

The requirements for accuracy and reproducibility are reasonably high for this type of calorimetry. Inaccuracy limits should not exceed one percent.

Because many of the above characteristics had been achieved separately by previous microcalorimeters, the design and construction of a microcalorimeter combining these requirements was feasible.

A calorimeter has been realized and its application to a series of biological problems is described in this paper. Several aspects of the work have subsequently been expanded and are being reported elsewhere in greater detail.

CALORIMETER

Selection Criteria

The microcalorimeter is of the heat-conduction or Tian type rather than the adiabatic type or the isoperibol type which depend on the measurement of a temperature rise, because of the success of such calorimeters in the hands of Tian and Calvet [1], Benzinger [2], Evans [3], Wadsö [4], and others in obtaining greater speed than the other types and high precision in measuring small quantities of heat.

The calorimeter is of the batch type, used by most of the previous investigators, rather than the flow (or stopped-flow) microcalorimeter such as have been successfully developed by Pennington, Brown, Berger and Evans [5], Wadso [4], and others.

While the latter have better promise of achieving faster response times than the batch type, they have not been investigated by as many laboratories and they require more reactant materials in their flow lines than the batch type, which has no flow lines. The batch type was also chosen because of ease of separation of sequential samples, conservation of materials, and the longer history of development.

Design Parameters and Theory

A heat-conduction microcalorimeter consists of a calorimeter compartment or cell surrounded by a heat-flow device which is in turn surrounded by a heat sink or block. The block is protected from environmental temperature changes by one or more constant-temperature jackets (Tian [6] put his whole calorimeter underground). The heat-flow device consists of multiple-junction series-connected thermocouples between an inner face in good contact with the calorimeter and an outer face in good contact with the block. The general features of a heat-conduction calorimeter are illustrated in figure 1, which also illustrates the design features discussed below, and incorporated in the NBS Mark I microcalorimeter.

Calibration Constant or Sensitivity

When heat is generated in the calorimeter (by electrical energy or by chemical energy) the inner face of the heat-flow device becomes warmer than the outer face and the heat which flows across is proportional to the output voltage. The proportionality constant is called the calibration constant, F, of the calorimeter. $F = h/\epsilon_0$, where h is the thermal conductivity of a single thermocouple pair, and ϵ_0 is the thermoelectric coefficient of the thermocouple material.

The materials of which the thermocouple is made and h_o (the thermal conductivity of the materials) fix the values of ε_o. Since $h = h_o a/\ell$, the value of the calibration constant F, thus depends on the length (ℓ) and cross sectional areas (a) of the thermocouple materials. Note that the calibration constant F does not depend on the number of thermocouple junctions.

The value of F, the calibration constant, determines the thermopile sensitivity of the calorimeter. That is 1/F gives the volts output per watt of power input into the calorimeter. In the calorimeter described here, using commercially available solid-state thermocouples N and P type bismuth sellenide-bismuth telluride-bismuth antimonide (ε_o = 400 $\mu V \cdot K^{-1}$), the calibration constant F = 16.7 $W \cdot V^{-1}$, or 1/F = 0.060 $V \cdot W^{-1}$ = 60 $mV \cdot W^{-1}$.

Heat Flow Path

The number of thermocouples is important in order to (a) cover the surface of the calorimeter uniformly such that all or most of the heat conduction is through the thermocouples or through parallel insulation or rods which are the same for each thermocouple; and to (b) obtain the desired time constant for the calorimeter.

If the temperature of the outer surface of the calorimeter cell is non-uniform due to non-even generation of heat in various parts of the cell and is not smoothed out at the surface then it is important that each thermocouple (including the insulation or rods paralleling each thermocouple) have the same heat conductivity so that the total heat flow will be properly accounted for.

The rate of heat flow, w_i, through a single thermocouple path is derived as follows:

$$w_i = h_i (\theta_c - \theta_b)_i$$

$$v_i = \varepsilon_{oi} (\theta_c - \theta_b)_i$$

$$f_i = h_i / \varepsilon_{oi}$$

$$\therefore w_i = f_i v_i$$

where h_i is the thermal conductivity of the ith path, $(\theta_c - \theta_b)_i$ the temperature difference between the calorimeter surface and block at that point, v_i the voltage generated by that thermocouple, ε_{oi} the thermoelectric constant for that thermocouple, and f_i the calibration constant for that path.

The total heat flow from the calorimeter is:

$$W = \Sigma w_i = \Sigma f_i v_i$$

If all the f_i are equal and the thermocouples are connected in series, we have

$$V = \Sigma v_i$$

$$F = f_i = \text{constant}$$

$$W = FV$$

in which F is defined as before. It is not difficult to obtain uniform thermocouple material so that the ϵ_{oi} are equal, but care must be exercised so that soldering does not alter the composition differently on each thermocouple. However, it is more difficult to get the h_i equal because the path lengths are small.

We can allow for a non-uniform temperature distribution if we can make the f_i equal. However, it is not so critical to keep the f_i equal as it is to make the heat flow to the surface in the same way in each experiment so that the temperature profile of the surface is the same in each experiment. This is done in the present calorimeter by using a silver or copper box to contain the calorimeter cell, while attempting to keep the f_i equal also. It is not necessary to have a very large number of thermocouples for a microcalorimeter if the calorimeter cell is small.

<u>Time Constant</u>

If heat is generated almost instantaneously in the calorimeter, the voltage V rises quickly above the baseline voltage, reaches a peak, and then exponentially decays back to the baseline voltage. If constant electrical power supplied to the calorimeter is suddenly turned off, the heat flow rate and V will exponentially decay to zero. It is this exponential decay with its time constant τ which largely determines the length of time an experiment will take in experiments where we wish to measure the total heat.

The voltage decay will to a first approximation follow the well known relationship:

$$V = V_o \exp (-t/\tau)$$

where $\tau = C_p/nh$, the time constant of the calorimeter. Since Q, the total heat, is $F\int V dt$, if we wish to measure the heat produced in the calorimeter with an accuracy of 0.1%, the voltage must decay to 0.1% of its maximum value. This will take a period of 6.9 τ or 10.0 half-times. Thus if we want this to be 5 minutes, we will want τ to be 43.4 seconds or the half-time to be 30.1 seconds.

To make τ small we make Cp, the heat capacity of the calorimeter and contents, as small as is practical and nh, the conduction to the block, as large as is practical, remembering that we want h to be small to obtain a useful sensitivity or calibration constant F (see above). Thus a large number of thermocouples, n, is helpful.

The time constant is calculated above on the assumption that the calorimeter and contents are at a uniform temperature. This is a reasonable assumption if the materials have high thermal diffusivity; under this condition thermal equilibrium can be obtained quickly in relation to the time constant. Because the thermal diffusivity of water is not high it is important that the thickness of the water or solution be small (of the order of a few mm) in order to keep the time constant small. In the calorimeter described the thickness of solution is 2.5 mm with 0.5 mm plastic walls making 1.75 mm as the maximum path length.

Block Temperature

Another factor affecting the precision and accuracy and the limiting sensitivity of the calorimeter is the constancy of temperature of the block. The aluminum block should be large enough so that the small amounts of heat introduced into the calorimeter cell will not significantly change the block temperature. An aluminum cylinder about 12 cm in diameter by about 12 cm long is large enough for the present case.

The temperature of this block must also be kept constant to about one microkelvin. This constancy cannot now be obtained by an "active" control such as a heater and detector on the block nor by passing controlled-temperature air over the block since a µK is difficult to detect and more difficult to actively control in a large block. It can be obtained by use of a "passive" control or "nesting". That is, the block can be suspended in stationary air in a constant temperature jacket. Using this arrangement the jacket must be controlled to 0.001 K and should not drift more than this over the length of any experiment. In the present calorimeter the jacket is also nested in another jacket with about a 5 cm annular space filled with insulation between them. Control of the jacket to about 0.0001 K was achieved in this way. With this degree of control, twin cell operation offers no advantage.

Speed of Operation

In order to perform many experiments per day, it is necessary to load and unload the calorimeter cell quickly without disturbing the block appreciably. This was made possible by using small plastic cells which slip into a silver box permanently stationed between the heat-flow discs. Removal of a cell and the insertion of a new cell is almost entirely equivalent to a heat process in a cell already in the silver box. Thus little time is lost between experiments.

Heat Transfer in the Calorimeter

A loose "push" fit allows sufficient conductivity between the plastic cells and the silver box so that the time constant is not adversely affected. The variations of mass or heat capacity of the cell and contents do not affect the measurement of the amount of heat significantly and only affect the time constant by a small amount in the usual case.

Since silver has a very good thermal diffusivity, the box serves to spread the heat generated in the plastic cell uniformly among the various thermocouple faces of the heat-flow disc; it thermally short-circuits the two heat-flow discs and averages their responses; and it provides for a heat-flow path which is very similar from one experiment to another. Thus differences in the thermocouple elements or in the two heat-flow discs are smoothed out and it is not critical that these be identical.

The plastic cells (figure 2) are 2 cm x 2 cm x 0.35 cm outside dimensions. These slip into a slot in the silver box which is permanently cemented to the flat heat-flow discs. This leaves the four narrow edges not in contact with the heat-flow discs but separated by air from the block. As the conductivity of air is about 0.02 that of the heat flow discs and the area exposed is about 0.25 of the total area, about 0.005 of the heat will pass to the block through the air. This heat will be proportional to the temperature difference between the cell and the block. To the degree this is true, the heat from the edges will be accounted for completely. The same effect is present in the electrical calibration experiments as in the reaction experiments and thus is compensated by the calibration. Since the plastic cell sits in the silver box, the top of the cell is left uncovered. Only this edge deviates appreciably from the required temperature homogenity. The top comprises about 1/15 of the total area and its temperature may be slightly warmer than the silver box. Assuming an error of half of the excess heat losses from this edge, the error should be less than 1/1500 on this account.

Electrical Heater

A bifilarly wound electrical heater is built into the silver box. The heater and its leads are permanently in place and any heat they conduct to the block from the silver box is largely accounted for in the calibration. The heat transferred will again be proportional to the temperature difference between the calorimeter block and their station at the temperature of the silver box.

INSTRUMENTATION, TESTING AND CALIBRATION OF THE
NBS MARK I MICROCALORIMETER [7]

Measurement of Thermopile Voltage

The thermopile voltage was amplified using a linear D.C. amplifier with the gain generally set at 10^5 and the rise time at 5 seconds. It was not found necessary to use either the amplifier's zero suppression or input filter. Electrical ground for the thermopile signal was established at the input to the amplifier.

Calibration of the amplifier was periodically performed in the following manner: current from a stable D.C. power supply was passed through four standard resistors (1-Ω, 10-Ω, 100-Ω, and 10-kΩ) in series. The input to the amplifier was the voltage across one of the lower valued resistors. From the voltage measured across both the entire voltage divider network and the amplifier's output terminal adjusted for the amplifier zero, the gain of the amplifier was calculated. The amplifier gain, as measured in this manner, was found to be within 0.02% of its nominal value on the 10^5 scale and stable to within $\pm$0.006% over a period of one year.

The amplifier output was recorded both in strip chart and digital form. The data logging system is shown schematically in figure 3. The analog to digital conversion was accomplished using an autoranging digital voltmeter with least readings of 1 μv or 10 μv. The stability and accuracy of this voltmeter were checked periodically against saturated standard cells calibrated in the NBS Electricity Division. The digital voltmeter was found to be accurate within 0.0025% at a reading of 1.018 v, and to be stable to $\pm$0.001% over a period of one year.

Calculation of Areas

A calorimetric measurement consisted of measuring the thermopile voltage as a function of time continuously (or at 3 second intervals). These readings may be divided into three periods: a fore period, a main period, and an after period. During the fore period, no process heat is being liberated in the reaction vessel. Ideally, the mean of the measured voltages would be zero and flucuations about the mean would be due only to Johnson noise. The main period begins with the initiation of the chemical reaction or electrical heating and the accompanying heat effect, which is detectable as a heat flow across the thermopile and a corresponding change in the thermopile voltage. After completion of the reaction or electrical heating, there is an exponential decay of the voltage to the after period baseline, which ideally should be the same as the fore period baseline.

The total heat effect, Q, to be associated with electrical energy or chemical reaction is proportional to the area, A, of the plot of voltage readings, E, versus time, t.

$$Q = F \cdot A$$

This area in millivolt-seconds (mV·s) is the area of the curve above the extrapolated baseline from time t_1 (at the end of the fore period) to time t_2 (at the beginning of the after period).

To obtain the extrapolated baseline during the reaction period, a straight line was assumed from $E(t_1)$, the voltage at the end of the fore period, to $E(t_2)$, the voltage at the beginning of the after period. $E(t_1)$ and $E(t_2)$ were obtained by a linear least-squares fit of the fore and after period readings, respectively. The area was obtained by the use of the trapezoidal rule.

All of the above calculations were done on a digital computer utilizing the digital voltage and time readings, taken at 3-second intervals throughout the experiment.

Temperature of the Calorimeter

The temperature of the microcalorimeter was controlled by control of the current through a heater wound on the inner jacket of the calorimeter. The regulator possessed proportional, rate, and reset control modes and these were adjusted to obtain the optimum in control response consistent with the absence of oscillations. A modified quartz oscillator thermometer [8] taped to the inside surface of the inner jacket was used to check the temperature stability of the inner jacket. A temperature record (readability +10 µK) obtained over a 15 hour period indicated a total drift of 0.00094 K for this entire period, with short term temperature fluctuations less than 0.0002 K.

The absolute temperature of the inner jacket was measured with a calibrated mercury-in-glass thermometer and a copper-constantan thermocouple junctioned between the inner jacket of the microcalorimeter and the mercury-in-glass thermometer.

Time Constant and Speed of Operation of Calorimeter

The half-response time of the microcalorimeter, as determined by observation of the exponentially decaying portion of reaction peaks, was observed to be 30 to 40 seconds. This corresponds to a time constant of about 50 seconds. A wait of twenty half-response times, or 10 minutes, is adequate to insure proper thermal equilibration of a calorimetric cell in the calorimeter after transfer from an external pre-equilibration vessel kept within a degree of the calorimeter temperature. The amount of time required to complete a run, following the thermal equilibration period in the calorimeter, is 15 to 20 minutes. This allows sufficient time for both a mix and remix operation with the necessary recording of the fore, main, and after period voltage readings for each. Thus, total time elapsed from the inserting a cell into the calorimeter to end of a run is about 25 to 35 minutes. Using these methods it is possible to perform about 15 separate experiments during the course of an 8 hour day.

Quality of Baseline

The quality of the baseline or thermopile voltage readings, is an important characteristic of a heat conduction microcalorimeter, and is determined by the absence of electrical or thermal interferences and is ultimately limited by Johnson noise. For measurement of the heat effects accompanying rapid processes it is necessary to be concerned about the quality of the baseline over only a 15 to 20 minute interval as a consequence of the rapid response time of the calorimeter. Peak-to-peak noise was generally within 30 nV over this time interval. Since the sensitivity of the instrument is 60 μV· mW^{-1}, this corresponds to 0.5 μW of thermal power fluctuations. Average baselines well within these fluctuations could be drawn. The long term drift of the baseline has been found to be typically less than 100 nV over a 24 hour period. The calorimeter response to a signal near its lower limit of usefulness is shown in figure 4. The measured value of the thermopile voltage was generally within 0.5 μV of electrical zero. Large deviations from electrical zero indicate experimental error with such causes as: (1) water leaks from the reaction cells; (2) a severe thermal disturbance of the calorimeter, such as loss of temperature control, and (3) an effect believed to be due to a pressure disturbance of the thermopiles whenever a reaction cell was fit too tightly into the silver box of the microcalorimeter. This latter effect was observed, for example, when excess wax, used in sealing some cells, was left on the side of that cell.

Electrical Calibrations

Electrical calibrations of the microcalorimeter were performed using the permanent resistance heater mounted in a cavity of the silver box and measurement procedures previously described [8]. Power input to the calorimeter was calculated as the product of the potential across the heater and the current thru it, with a small correction (0.005%) applied for power generated in the heater leads. In applying this correction, it was assumed that one-half the power generated in the heater leads could be apportioned to the jacket (i.e. the aluminum block) and the other half to the calorimeter vessel.

The results of electrical calibrations are given in Tables 1 and 2. The results, shown in Table 1 were obtained when the calorimeter was at a temperature of 29.85°C and were performed (with the exception of run no. 25) with no sample cell contained in the silver box of the microcalorimeter. In Table 2, the results reported were taken at an assigned calorimeter temperature of 30.80°C, and a sample cell containing approximately 0.3 ml of water was in the silver box. The calibration constant (reciprocal sensitivity, in units of $W \cdot V^{-1}$ or $J \cdot V^{-1} \cdot s^{-1}$) was calculated as the ratio of the heat input in a measured time interval (mJ) to the measured peak area (units of $mV \cdot s$). In a few experiments the calibration constant was obtained from the steady state deflection (V) obtained after application of a steady state power for a long time (13 to 16 half-response times of the calorimeter). The values of the calibration constant, obtained in this manner, are given in parenthesis in Tables 1 and 2.

It will be noted that the calibration constant is the same, within the imprecision of the measurements, whether we use the integral energies or the steady power-deflection method (values in parentheses). The integral energy results are used since the areas were determined in the same way as in chemical reaction experiments. The calibration constant is thus taken as 16.67 $W \cdot V^{-1}$ at either temperature. Inspection of the data in Tables 1 and 2 shows no large trend with power level from 0.01 to 2.00 mW, time of heating from 10 to 800 seconds, or energy from 1 to 1400 mJ. Four runs were omitted in averaging. The imprecision of the determinations of the calibration constant with electrical energy is approximately 0.4 percent for energies from 1 to 1400 mJ.

We are indebted to H. Suga for pointing out a possible systematic trend of the calibration data with total energy, which is most evident at low energies.

Procedures Used in Reaction Calorimetric Experiments

For the calibration and testing of the calorimeter, a "wet cell" procedure was used in all experiments. In this procedure, a small quantity of water was permitted to remain on the interior walls of the vessel to facilitate mixing of the solution. This was replaced by a "dry cell" procedure using cells having a hydrophilic surface treatment for the biochemical experiments.

Solutions were introduced into the reaction cells with syringes. Quantities of solutions introduced into a reaction cell were determined by weighing the cells.

Cell sealing was done in one of two ways: (1) use of small Kel-F or Teflon plugs that fit into the cell entrance portals and were sealed over with melted wax and (2) use of no. 2-1 Buna-N "O"-rings that were compressed by means of small Teflon plugs. The sealing process is extremely important, for if a cell is not properly sealed, the vaporization of water, will cause a large endothermic heat effect which will dwarf the heat effect being measured and make its determination uncertain.

Table 1. Results of Electrical Calibrations (Series I). Temperature is 29.85°C.

Run No.	Power Introduced (mW)	Time of Heating (s)	Heat Input (mJ)
1	0.010019	20.017	0.2006
2	0.010019	600.125	6.0128
3	0.010019	100.254	1.0045
4	0.020037	10.457	0.2095
5	0.020037	600.543	12.0332
6	0.020037	100.226	2.0083
7	0.050117	10.282	0.5153
8	0.050117	599.925	30.0665
9	0.050117	100.147	5.0191
10	0.099820	10.228	1.0210
11	0.099820	600.342	59.9263
12	0.099820	100.227	10.0047
13	0.197226	10.379	2.0470
14	0.197226	605.393	119.399
15	0.197226	100.263	19.7745
16	0.498069	10.285	5.1226
17	0.498069	617.153	307.385
18	0.498069	99.949	49.7815
19	0.998949	10.311	10.3002
20	0.998949	600.808	600.177
21	0.998949	100.151	100.046
22	1.99239	10.304	20.5296
23	1.99239	600.194	1195.82
24	1.99239	100.188	199.614
25	0.99901	650.401	649.754

Area (mV·s)	Steady State Deflection (μV)	Calibration Constant (W·V^{-1})
0.0118	---	*16.996
0.3626	0.6066	16.582 (16.517)
0.0607	---	16.548
0.0144	---	*14.551
0.7226	1.2135	16.653 (16.512)
0.1211	---	16.583
0.0323	---	*15.954
1.7993	2.991	16.710 (16.756)
0.2998	---	16.741
0.0610	---	16.737
3.5866	5.986	16.708 (16.676)
0.5985	---	16.716
0.1223	---	16.738
7.1509	11.840	16.697 (16.658)
1.1844	---	16.696
0.3066	---	16.708
18.4100	29.820	16.697 (16.703)
2.9821	---	16.693
0.6185	---	16.654
35.9621	59.824	16.689 (16.698)
5.9948	---	16.689
1.2304	---	16.685
71.6537	119.32	16.689 (16.698)
11.965	---	16.683
38.8935	59.783	16.706 (16.711)

Mean calibration constant, W·V^{-1} 16.682 (16.659)

Average deviation, W·V^{-1} ±0.035 (±0.21%)

*Omitted

Table 2.   Results of Electrical Calibrations (Series II).   Temperature is 30.80°C.

Run No.	Power Introduced (mW)	Time of Heating (s)	Heat Input (mJ)
1	0.49728	10.979	5.4596
2	0.49728	11.570	5.7535
3	0.49728	10.227	5.0857
4	0.49740	9.981	4.9645
5	0.49736	20.428	10.1601
6	0.099715	10.220	1.0191
7	0.099715	100.197	9.9912
8	0.099715	755.170	75.3019
9	0.49761	10.457	5.2035
10	0.49761	100.210	49.8652
11	0.49761	699.751	348.201
12	0.99806	10.477	10.4566
13	0.99806	100.256	100.061
14	0.99806	809.019	807.446
15	1.99062	10.201	20.3063
16	1.99062	100.117	199.294
17	1.99062	729.464	1452.08

Area ($mV \cdot s$)	Steady State Deflection (μV)	Calibration Constant ($W \cdot V^{-1}$)
0.3263	---	16.732
0.3492	---	16.476
0.3021	---	16.835
0.2971	---	16.710
0.6082	---	16.705
0.0593	---	*17.185
0.5987	---	16.688
4.5161	6.039	16.674 (16.512)
0.3164	---	16.446
2.9798	---	16.734
20.890	29.79	16.668 (16.704)
0.6309	---	16.574
5.9993	---	16.679
48.433	59.834	16.671 (16.680)
1.2276	---	16.541
11.954	---	16.672
87.119	119.380	16.668 (16.675)

Mean calibration constant, $W \cdot V^{-1}$ 16.655 (16.643)

Average deviation, $W \cdot V^{-1}$ ±0.072 (±0.43%)

*Omitted

Table 3.   Blank Heats of Mixing

A.   Simple Mechanical Rotation   1

Run. No.	Area (mV·s)
1	−0.0020
2	−0.0038
3	−0.0006
4	−0.0021
5	−0.0015
6	−0.0045
7	−0.0020

Average Area = −0.0024 ± 0.0011 mV·s

$$Q = -40 \pm 18 \ \mu J$$

B.   Water vs Water

Run. No.	Area (mV·s) 1st Rotation	2nd Rotation
1	+0.0114	+0.0045
2	+0.0056	+0.0035
3	+0.0150	+0.0000
4	+0.0075	+0.0009
5	+0.0104	−0.0013

1st Rotation: Average Area = +0.0100 ± 0.0028 mV·s

$$Q = +167 \pm 47 \ \mu J$$

2nd Rotation: Average Area = +0.0015 ± 0.0020 mV·s

$$Q = +25 \pm 33 \ \mu J$$

C. <u>0.08 N NaOH vs 0.08 N NaOH (Wet-cell)</u>

Run.No.	Area (mV·s)	
	1st Rotation	2nd Rotation
1	+0.0658	−0.0006
2	+0.0710	+0.0009
3	+0.0782	−0.0004
4	+0.0719	−0.0004
5	+0.0568	+0.0020

1st Rotation: Average Area = +0.0687 $\pm$ 0.0060 mV·s

$\qquad\qquad\qquad$ Q = +1147 $\pm$ 100 µJ

2nd Rotation: Average Area = +0.0003 $\pm$ 0.0010 mV·s

$\qquad\qquad\qquad$ Q = +5 $\pm$ 17 µJ

D. <u>0.01 N HCl vs 0.01 N HCl (Wet-cell)</u>

Run. No.	Area (mV·s)	
	1st Rotation	2nd Rotation
1	+0.0010	−0.0004
2	+0.0046	−0.0025
3	+0.0183	−0.0039
4	+0.0227	−0.0094

1st Rotation: Average Area = +0.0117 $\pm$ 0.0089 mV·s

$\qquad\qquad\qquad$ Q = +195 $\pm$ 148 µJ

2nd Rotation: Average Area = −0.0041 $\pm$ 0.0027 mV·s

$\qquad\qquad\qquad$ Q = −68 $\pm$ 45 µJ

Following the sealing of a cell, it was then put into an aluminum-block pre-equilibrator. Following a five to ten minute wait, the cell was then transferred to the micro-calorimeter for final equilibration. This generally took about 15 minutes. The reaction (main period) was then initiated by a 360-degree rotation of the microcalorimeter block and 360-degrees return. After an adequate number of data points had been accumulated in the after period, a remix was performed to ascertain the completeness of reaction. The cell was then removed from the calorimeter and the next run was performed.

Blank Heat Effects

A factor which effects the precision and accuracy of the microcalorimeter is reproducibility of the heat effect accompanying the necessary experimental operation of reaction initiation by mixing. The results of measurements of the magnitude of this heat effect are given in Table 3. A simple operation of mechanical rotation of the calorimeter block yields an apparent endothermic heat effect of -40 + 18 μJ (average deviation), (Table 3A). The cause of this effect is not fully understood and may be due to movement of wires through the Earth's magnetic field during rotation.

When solutions are present in the cell the heat effect accompanying the first rotation is always larger than that accompanying the second rotation or remix. In the case of the mixing of water with water (Table 3B), this may be attributable to wetting of the cell or to some other surface interaction. However, in the case when sodium hydroxide (Table 3C) and hydrochloric acid (Table 3D) are mixed with themselves, the effect is a consequence of the "wet cell" procedure used and was the result of the dilution of acid or of base by the pure water on the walls above the solutions in the cell.

Acid-Base Test Reaction

The enthalpy of neutralization of hydrochloric acid with sodium hydroxide was used as a test of the accuracy of the microcalorimeter. The hydrochloric acid used in these experiments, kindly provided by George Marinenko of the Analytical Chemistry Division, had been analyzed by a coulometric procedure [9, 10]. This stock acid, as received, had a concentration of 0.199242 meq·g^{-1} which was believed to be accurate to within +0.02%. This stock was used in the preparation of more dilute acid solutions by weighing. A saturated stock solution of sodium hydroxide was prepared using Baker analyzed sodium hydroxide pellets and distilled water. This solution was allowed to stand for several weeks to permit precipitation of sodium carbonate; then a portion of the clear solution was withdrawn from the saturated sodium hydroxide solution and diluted using freshly boiled distilled water. The concentration of this sodium hydroxide solution was determined by titration against potassium acid phthalate (NBS Standard Reference Material No. 84g). In all calorimetric runs, the sodium hydroxide was used in substantial excess of stoichiometric.

The results of calorimetric measurements of the heat of neutralization of hydrochloric acid with sodium hydroxide are given in Table 4. In analyzing these results, a calibration constant of 16.67 $J \cdot V^{-1} \cdot s^{-1}$ was used. The measured heat (Q), given in Table 4, includes a correction of -1147 µJ, for the blank heat of mixing as determined when 0.08 N sodium hydroxide was mixed with itself. Corrections for heats of dilution to the standard state [11] were applied using tabulated data [12] for the relative apparent molal enthalpy and heat capacity of the reactants and products. Also a correction was applied [12] to convert the data to 25°C. No correction was applied for the change in vapor pressure of water above the solutions. The total estimated effect was ∿0.02%.

The final value, $\Delta U°$, is the internal energy change of ionization of water at 25°C. Since $(\Delta PV)°$ for this process is small $\Delta U°$ is taken as equal to $\Delta H°$, the enthalpy of ionization of water.

The resulting values of $\Delta U°$, given in Table 4 were averaged (with the exceptions of four runs to yield a value of $\Delta U° = -55.74$ $kJ \cdot mol^{-1}$ with an average deviation of 0.33 $kJ \cdot mol^{-1}$. The four runs discarded, out of a total of 22 measurements, all differed from the average of the remaining results by at least a factor of seven times the average deviation. The low results of these four runs are presumed to be due to some pre-reaction of the solutions. The result agrees, within the experimental imprecision, with the "best" literature value of -55.835 ± 0.104 $kJ \cdot mol^{-1}$ [12].

The imprecision of measurement of the heat of a chemical reaction is thus about 0.6 percent when the heat is of the magnitude of 50 mJ or more. In the present experiments it was necessary to make a 1 mJ correction for the blank mixing effect caused by the "wet cell" procedure.

Table 4. Results of Acid-Base Calorimetric Experiments

Run No.	Temperature (°C)	Mass of acid (g in air)	Mass of base (g in air)	Acid concentration (mol kg^{-1})	Mix Area (mV·s)
1	29.85	0.101753	0.100950	0.010322	3.5720
2	29.85	0.100650	0.100550	0.010332	3.5577
3	29.85	0.094590	0.101715	0.010332	3.3740
4	29.85	0.100865	0.101550	0.010332	3.4084
5	29.85	0.095260	0.102130	0.010332	3.4000
6	29.85	0. 06380	0.104790	0.010577	3.8102
7	29.85	0.101067	0.102005	0.010577	3.6311
8	29.85	0.102465	0.102143	0.010577	3.7063
9	30.80	0.145795	0.153165	0.019669	9.6496
10	30.80	0.149850	0.150615	0.019669	9.8576
11	30.80	0.149341	0.153744	0.019669	8.0123
12	30.80	0.149345	0.151745	0.0099243	4.9772
13	30.80	0.152340	0.104840	0.0053055	2.4034
14	30.80	0.149549	0.101190	0.0053055	2.7310
15	30.80	0.145230	0.101675	0.0053055	2.6529
16	30.80	0.149160	0.100730	0.0097799	4.9209
17	30.80	0.150840	0.101715	0.0097799	4.9828
18	30.80	0.148924	0.102438	0.0097799	4.9785
19	30.80	0.150645	0.101251	0.0097799	4.9776
20	30.80	0.147990	0.102026	0.0097799	4.9199
21	30.80	0.148834	0.102352	0.0097799	7.6777
22	30.80	0.150233	0.102176	0.0097799	9.9808

Base concentration is 0.080856 mol kg^{-1} in all experiments.

* Omitted

** Corrected by -1.147 mJ for "blank" mixing.

Remix Area (mV·s)	Heat** (mJ)	$-\Delta U$ (kJ mol^{-1})	Correction Std. State (kJ mol^{-1})	Correction to 25°C (kJ mol^{-1})	$-\Delta U°$ (kJ mol^{-1})
+0.0003	58.505	55.592	-1.195	1.040	55.437
+0.0004	58.268	55.973	-1.198	1.040	55.816
-0.0001	55.192	56.415	-1.228	1.040	56.227
+0.0081	55.903	53.587	-1.201	1.040	*53.426
-0.0053	55.539	56.371	-1.182	1.040	56.184
+0.0107	62.657	55.628	-1.172	1.040	55.497
-0.0033	59.432	55.539	-1.182	1.040	55.398
+0.0031	60.795	56.037	-1.176	1.040	55.901
+0.0003	159.81	55.670	-0.859	1.234	56.045
+0.0006	163.28	55.341	-0.848	1.234	55.727
+0.0003	132.50	45.067	-0.854	1.234	*45.443
+0.0007	81.880	55.187	-1.311	1.234	55.110
+0.2595	43.266	53.476	-1.857	1.234	*52.852
-0.0050	44.319	55.802	-1.841	1.234	55.195
-0.0037	43.038	55.798	-1.874	1.234	55.158
+0.0005	80.938	55.426	-1.114	1.234	55.545
-0.0004	81.956	55.498	-1.113	1.234	55.618
+0.0015	81.915	56.184	-1.124	1.234	56.294
+0.0160	82.142	55.696	-1.112	1.234	55.818
+0.0004	8.920	55.852	-1.125	1.234	55.960
-0.0239	126.51	43.196	-0.748	1.234	*43.681
⊢0.0013	165.31	55.916	-0.745	1.234	56.404

Mean of $-\Delta U°$ at 25°C, kJ·mol^{-1} 55.74

Average deviation, kJ·mol^{-1} ±0.33
(±0.59%)

APPLICATIONS OF MICROCALORIMETRY TO CLINICAL LABORATORY PROBLEMS

Assay of Substrates in Complex Fluids--General Principle

To use the microcalorimeter in clinical measurements a necessary first step is Proof of Principle. To perform this proof requires a demonstration on a well-known substance that the calorimetric procedure works well at normal and clinically important concentrations in buffered aqueous medium and in real biological fluids (i.e. serum).

The problem, of course, is to contend with the fact that a biological fluid typically consists of a complex mixture and the substrate of interest must be identified quantitatively in the presence of this complex mixture of considerably variable composition. The only procedure offering a reasonable hope of accomplishing this is to apply the selective reactivity of the desired substrate in the presence of a specific enzyme.

For an adequate treatment of any particular substrate-enzyme process the following chemical problems must be under control:

(a) Precise knowledge of the reaction occurring
(b) Energetics of the reaction
(c) Kinetics of the reaction
(d) Absence of complicating reactions, parallel or sequential.

By far the most controversial factor is the demonstration that competing or sequential reactions are not occurring in significant amounts. Absolute certainty on this point may be beyond our present science. Probably the best assurance within reach is to be quite sure of the thermochemistry of the process of interest, so that deviations will be clearly evident.

A very important chemical problem is the enzyme technology without which the technique would have been impossible. The major factors having a bearing on the enzyme-specific procedure are:

(a) The right enzymes for the job
(b) Availability of sufficient pure enzymes.

The rate of reaction will determine (1) the time scale for the reaction occurring in the calorimeter, and (2) the degree of completeness of the reaction after a given elapsed time. The reaction rate of an idealized enzyme catalyzed process is illustrated in figure 5. In the early stages of the reaction the substrate may be assumed to have a concentration sufficient for all sites on the enzyme to be occupied at all times. In that circumstance the rate of reaction is constant and is determined by the concentration of enzyme.

274

As substrate is depleted by the reaction, a region is reached in which the substrate concentration is no longer sufficient to saturate the enzyme, and the reaction rate will diminish with time, going to zero rate as the substrate vanishes. As a result to achieve rapid reaction and assure essentially complete reaction, the enzyme must be used in comparatively large amounts.

The magnitude of substrate concentrations to be dealt with in specific instances are suggested by the list of normal concentrations in serum given in table 5.

Table 5

Normal Concentrations in Serum

		$mg/100 \ cm^3$
Cholesterol		140-250
Glucose		70-110
Creatinine		0.7-1.5
Uric Acid	(male)	3.8-7.1
	(female)	2.6-5.4
Lipids		360-765

To further place this problem in perspective, let us examine a typical reaction suitable for study: the glucose-oxidase catalyzed oxidation of glucose followed by catalase-catalyzed reduction of the peroxide formed. This is summarized in table 6. The energies given are only approximate, and the equations are a simplified statement of the overall process.

The first fact to observe is that the energy of reaction on a molar basis is not small, but rather is in the normal range of oxidative chemical reactions. The calorimetric sensitivity requirement is imposed by the small molar concentration of glucose (1 to 5 millimoles per litre) and by the need to use only a small volume of solution. For this typical example we find an expected energy change of about 45 millijoules, easily within the range of the microcalorimeter.

In any exploratory studies on the applicability of calorimetry to assay of biological fluids, this elementary exercise must be carried out, relating the expected molar energy change of prospective processes, the interesting concentrations of substrate, and the amount of fluid that may be required.

Table 6

Example - Glucose

(1) Glucose(aq) + O_2(aq) + $H_2O(\ell)$ ΔH kJ·mol^{-1}

     $\xrightarrow[\text{oxidase}]{\text{glucose}}$ gluconic acid(aq) + H_2O(aq)   -197

(a) H_2O_2(aq) $\xrightarrow{\text{catalase}}$ $H_2O(\ell)$ + $1/2\ O_2$(aq)   -100

Total O_2 consumed: 1/2 mol. Total Energy: -297

Consider a real solution in the normal concentration range

glucose	90 mg/100 cm^3	(900 mg/ℓ)
(M.W. 180)	5 mmol/ℓ	
Dilute by 5 $\longrightarrow$	1 mmol/ℓ	

In the calorimeter mix:

    150 μℓ glucose soln (150 nmol)

    150 μℓ enzyme in buffer

    ‾‾‾‾‾‾

    300 μℓ

O_2(dissolved; 0.25 mmol/ℓ) = 75 nmol

The energy of reaction:

    297 kJ·mol^{-1} x 150 nmol

    = 44.5 mJ (exothermic)

- - - - - - - -

Reaction medium:

    Saline phosphate buffer (pH = 6.0)

Enzyme

    Glucose oxidase w/ catalase impurity

      2000 units oxidase activity per 100 ml

Glucose Assay

Exploratory studies involving the batch calorimeter indicated that perhaps the glucose-glucose oxidase system was not the simplest system on which to begin the study. We, therefore, (on advice) shifted to the glucose-ATP reaction catalyzed by hexokinase.

The complete description of the work on the glucose-ATP-hexokinase reaction, is given in reference [13].

Glucose reacts with ATP in the presence of hexokinase according to reaction (1).

$$\text{Glucose (in buffer)} + \text{ATP (in buffer)} \xrightarrow[\text{Mg}^{+2}]{\text{hexokinase}}$$

$$\text{ADP (in buffer)} + \text{glucose-6-phosphate (in buffer)} \quad (1)$$

A proton produced in the above reaction in turn will react with the buffer present. The measured effect will be the sum of the enthalpy effects associated with these two reactions. The energy of protonation of the buffer is a significant fraction of the total energy measured (about 2/3). An additional important factor in obtained reliable calorimetric values is to prepare solutions in such a way as to avoid buffer mismatch in the substrate solution and the enzyme solution that are to be mixed.

The energies measured for reaction in aqueous medium, using measured amounts of glucose are given in figure 6. The line corresponds to $\Delta H = -61.4$ kJ$\cdot$mol^{-1}. The approximate concentrations of glucose are noted beside the line. The relationship is such that 0.5 mJ for 150 microlitres of solution corresponds to 1 mg glucose per deciliter of solution. The standard deviation of the points from the line is 0.31 mJ, which corresponds to less than one mg/dl. A small blank effect of mixing is observed as the intercept.

Using the energy-quantity relationship established in figure 6, six samples of human blood plasma and 45 samples of human blood serum have been assayed for glucose, using the glucose-ATP-hexokinase system. The results are shown in figure 7. For comparison, glucose was determined in the same samples at the NIH clinical laboratory in the autoanalyzer using glucose oxidase. The results are not strictly comparable because the autoanalyzer results are stated on a volumetric basis and the NBS results are stated on a mass basis. This effect is small.

Measurements at NBS were made usually in triplicate but sometimes in duplicate if the amount of samples was small. The mean average deviation of replicate measurements was 1.9 mg/100 g.

The time scale and general course of the energy measurement in the glucose determinations is shown in figure 8. In this figure the reaction was carried out in water, using a concentration of glucose somewhat less than normal for human blood serum. The remix indicates that reaction was essentially complete as a result of the original mixing process. For comparison we show also two superimposed measurements of glucose in human blood serum in figure 9. The time scale here is expanded about twofold. The duration is a characteristic of the calorimeter rather than of the reaction rate, which is such that reaction should be complete in about ten seconds.

Other Enzyme Catalyzed Processes

The energies of several enzyme catalyzed processes have been observed giving energies ranging from less than 30 to more than 100 $kJ \cdot mol^{-1}$. A list is presented in Table 7.

Table 7

Enzyme-Substrate	Buffer	pH
Hexokinase (glucose + ATP)	THMAM-HC1	7.6
LDH-(NADH + pyruvate)	Phosphate	6.5
CPK-(Creatine + ATP)	Glycine	9.0
NAD-ase-NAD	THMAM	9.9
Glucose oxidase-glucose	Phosphate	6.5
Uricase-uric acid	Borate-THMAM-HC1	9.2

Symbols used above:

ATP, adenosine-5'-triphosphate; LDH, lactate dehydrogenase; NADH, reduced nicotine adenine dinucleotide; CPK, creatine phosphokinase; NAD, nicotine adenine dinucleotide; THMAM, tris-hydroxymethyl aminomethane.

The numerical values observed were the results of one or two experiments in which conditions were not necessarily optimum-- and are qualitative information only.

Five of the above reactions were measured in this work, we were assisted by Dr. J. Everse of the University of California. The uricase-uric acid reaction was done in cooperation with N. Rehak and R. L. Berger, who have extended the work further at the National Institutes of Health.

The calorimeter response for the above reactions follows the same general time scale as with glucose indicating that the reactions proceed with reasonable velocity.

Bacterial Growth [14]

As an illustration of another class of potential applications of the microcalorimeter in the clinical laboratory, exploratory measurements were made of the growth patterns of the bacteria Enterobacter cloacae, Proteus rettgeri, and Klebsiella pneumoniae. (In this work we received the assistance of S. Farling and W. Russell of Instrumentation Laboratory, Inc., Lexington, Massachusetts, who supplied the bacteria and bacteriological techniques.)

The results of these experiments are very suggestive of the potential value of the technique, and indicate that the results found by previous investigators do not by any means exhaust the possible information to be obtained by bacterial studies. The subject of bacterial calorimetry was summarized by Forrest [15] (1969) who discussed the work of Prat, Belaich, Battley, Voivinet, and others, as well as his own work.

The upper part of figure 10 shows the growth pattern of Enterobacter cloacae. After an induction period and a period of thermal stabilization of the calorimeter an exponential-type growth begins which shows some structure Variations in heat production occur which may be due to some characteristic of the food supply or of the products of metabolism, and finally a fairly rapid termination of growth occurs. The lower part of the figure shows a repeat of the test on Enterobacter cloacae. We observe the same features. The second test has been somewhat displaced in time and in ordinate but not in scale. Note how closely the features of the two curves match.

Figure 11 shows a similar pair of curves for Proteus rettgeri. In this figure some differences of detail are observed in the second run which is superposed on the first run. Note the sharp break which occurs on one curve just after seven hours (on this time scale).

At this point it may be worth pointing out one feature of the instrumentation of our calorimeter for data logging. Although a recorder chart is made at the time of the experiment, the data are also taken as numerical values on a digital voltmeter and transferred to punched paper tape. The figures shown are actually transcribed by automatic data plotter from the numerical-data tape. This gives us the option of replotting the data on an expanded or a contracted scale, the better to observe certain features (see figure 12).

HEAT PRODUCTION OF ARBACIA PUNCTULATA EGGS [16]

The 1924 measurements of Rogers and Cole (17) were repeated in the present calorimeter, modified to use a larger (3 ml) reaction cell. Additional insulation was required around the calorimeter to allow it to be used successfully in an uncontrolled room. The fertilization efficiency was very sensitive to the manner of preparation of the mixing cell. Unfertilized eggs generate heat at the rate of about 0.3 nW/egg after one hour, fertilized eggs, at 0.6 nW/egg, with no burst of heat on fertilization. Fluctuations suggest a possible correlation with stages of development. The measurements were made using suspensions of about 10,000 eggs/ml. The rate of heat production may be sensitive to the concentration of eggs.

SUMMARY

(1) The microcalorimetric procedure for glucose assay using the hexokinase catalyzed phosphorylation reaction gives heat measurements proportional to glucose content in aqueous medium. When the measurements are extended to glucose in human serum and plasma the measured values correlate reasonably well with measurements by standard clinical procedures.

(2) Significant and rapid energy changes are observed in other enzyme catalyzed processes involving a variety of chemical mechanisms. From these observations a more general applicability of the enzyme-specific microcalorimetric assay method may be inferred.

(3) Observation of thermal patterns resulting from bacterial growth in culture shows characteristic curves reporducible for the same bacterium and distinguishable for different bacteria.

(4) Calorimeter development and analysis of calorimeter performance in this work show that the design of the calorimeter system plays an important role in determining the efficiency of the thermal recovery, the speed of the calorimetric measurements, and the amount of detail which are important for sensitivity, speed, and reliability are discussed.

(5) An important feature of the calorimetric procedure is its versatility. This results first from the facts that living organisms sustain themselves by chemical reaction processes and their typical constituents are thus chemically reactive; and second from the fact that chemical reactions are invariably accompanied by a thermal effect which forms the basis of a calorimetric measurement. These two characteristics complement one another with the result that calorimetry is applicable to a wide variety of processes associated with living organisms.

280

(6) Thermochemistry and its application via the technique of microcalorimetry is a procedure that is very promising for widespread application to the life sciences and in particular it has promise for application in clinical and biochemical laboratories.

This readiness results from the timely combination of recent improvements in microcalorimeter design, auxiliary calorimetric data logging and manipulation procedures, and enzyme technology, which now makes feasible procedures that have been discussed as possibilities for some years.

(7) In the application of this procedure to the specific needs of the clinical laboratory, several disciplines must be brought together, each of which plays an essential role. These are

(a) Calorimeter design and development.

(b) Fundamental thermochemistry of the processes involved in the biochemistry.

(c) Biochemical procedures in the various areas of potential applicability.

It is important that cross-discipline interaciton is essential for realization of the full benefits of calorimetry in clinical and biochemical laboratories.

References

[1] E. Calvet and H. Prat, "Recent Progress in Microcalori-metry", (The MacMillan Company, New York, 1963).

[2] T. H. Benzinger, Chapter in "A Laboratory Manual of Analytical Methods of Protein Chemistry", P. Alexander and H. P. Lundgren, eds., (Pergamon Press, Oxford and New York, 1969).

[3] W. J. Evans, Chapter in "Biochemical Microcalorimetry", H. D. Brown, ed., (Academic Press, New York, 1969).

[4] P. Monk and I. Wadsö, Acta. Chim. Scand. 22, 1842 (1968).

[5] S. N. Pennington, H. D. Brown, R. L. Berger, and W. J. Evans, Analytical Biochem. 32, 251 (1969).

[6] A. Tian, Bull. soc. chim. France 33, 427 (1923).

[7] E. J. Prosen and R. N. Goldberg, NBS Internal Report NBSIR 73-180 (1973).

[8] A. P. Brunetti, E. J. Prosen, and R. N. Goldberg, J. Research of the National Bureau of Standards, Vol. 77A, 599 (1973).

[9] G. K. Marinenko and J. K. Taylor, Anal. Chem. $\underline{40}$, 1645 (1968).

[10] G. K. Marinenko and C. E. Champion, J. Research NBS $\underline{75A}$, 421 (1971).

[11] C. E. Vanderzee and J. A. Swanson, J. Phys. Chem. $\underline{67}$, 2608 (1963).

[12] V. B. Parker, Thermal Properties of Aqueous Uni-univalent electrolytes, NSRDS-NBS-2 (U.S. Government Printing Office, Washington, DC, 1965).

[13] R. N. Goldberg, E. J. Prosen, B. R. Staples, R. N. Boyd, and G. T. Armstrong, NBSIR 73-178, April 1973.

[14] B. R. Staples, E. J. Prosen and R. N. Goldberg, National Bureau of Standards Internal Report NBSIR 73-181, April 1973.

[15] W. W. Forrest, Chapter VIII in Biochemical Micro-calorimetry, H. D. Brown, ed., (Academic Press, New York, 1969).

[16] E. J. Prosen and K. S. Cole, Abstract in Biological Bulletin $\underline{145}$, 450 (1973).

[17] C. G. Rogers and K. S. Cole, Biological Bulletin $\underline{49}$, 338 (1925).

Figure Captions

1. Conduction type microcalorimeter (NBS Mark I).

2. Plastic cell used for liquid samples. Sample and reagents are introduced by syringe through small pluggable openings. The liquids--on opposite sides of the partition are mixed by inverting the cell.

3. Data logging system for NBS microcalorimeter.

4. Response of microcalorimeter to 0.2 mJ electrical calibration heat. Heat supplied in 10 s.

5. Rate of enzyme catalyzed reactions (theoretical). V_{max} is the limiting rate determined by enzyme concentration in the period when substrate concentration is sufficient to saturate the enzyme. Near the origin when substrate concentration has dropped to a low value it no longer saturates the enzyme, and the law of mass action determines the rate.

6. Calibration reaction. Measured heat of the hexokinase-catalyzed phosphorylation of glucose, as a function of total glucose present.

7. Glucose concentration in human blood plasma and serum. Comparison of calorimetric results and autoanalyzer results.

8. Chart of calorimeter response for <u>hexokinase</u> catalyzed phosphorylation of glucose in aqueous buffer medium. The small signal labelled <u>remix</u> indicates that reaction was essentially complete on first mixing.

9. Chart of calorimeter response for <u>hexokinase</u> catalyzed phosphorylation of glucose in human blood serum. Two curves representing different concentrations of glucose are shown.

10. Repeated growth curves of <u>Enterobacter cloacae</u>. Curve B is displaced in time and in response but the scale sizes remain the same.

11. Repeated growth curves of <u>Proteus rettgeri</u>. The lower curve has been displaced in time and in response to indicate similarities. The scale sizes remain the same.

12. Magnification of initial portion of <u>Proteus rettgeri</u> growth curve. Note that rapid changes can be followed by the digital data acquisition system.

NBS Microcalorimeter

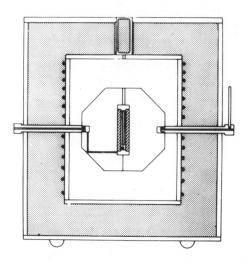

Fig. 1. Conduction type microcalorimeter (NBS Mark I).

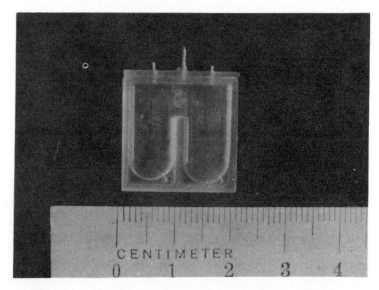

Fig. 2. Plastic cell used for liquid samples.
Sample and reagents are introduced by syringe
through small pluggable openings. The liquids--
on opposite sides of the partition are mixed by
inverting the cell.

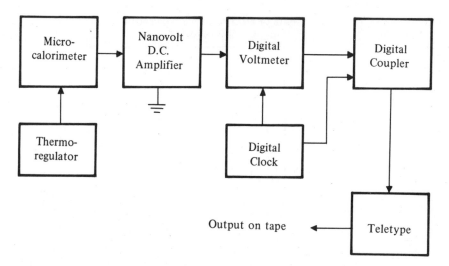

Fig. 3. Data logging system for NBS microcalorimeter.

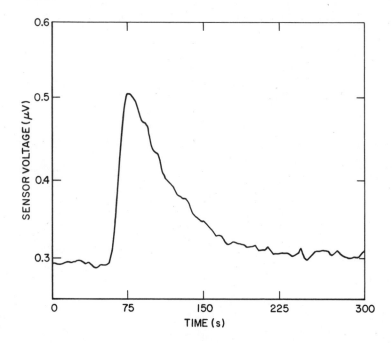

Fig. 4. Response of microcalorimeter
to 0.2 mJ electrical calibration heat.
Heat supplied in 10 s.

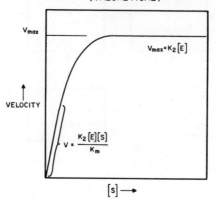

RATE OF ENZYME CATALYZED REACTION
(THEORETICAL)

V_{max}

$V_{max} = K_2[E]$

VELOCITY

$V = \dfrac{K_2[E][S]}{K_m}$

$[S] \longrightarrow$

Fig. 5. Rate of enzyme catalyzed reactions
(theoretical). V_{max} is the limiting rate deter
mined by enzyme concentration in the period
when substrate concentration is sufficient
to saturate the enzyme. Near the origin when
substrate concentration has dropped to a low
value it no longer saturates the enzyme, and
the law of mass action determines the rate.

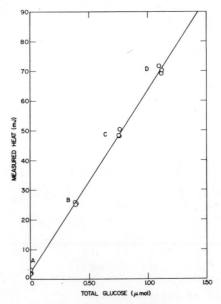

Fig. 6. Calibration reaction.
Measured heat of the hexokinase-
catalyzed phosphorylation of glucose,
as a function of total glucose present.

286

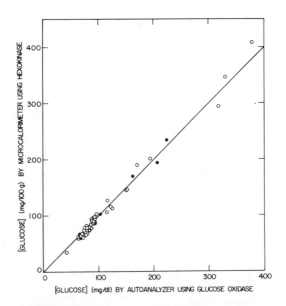

Fig. 7. Glucose concentration in human blood plasma and serum. Comparison of calorimetric results and autoanalyzer results.

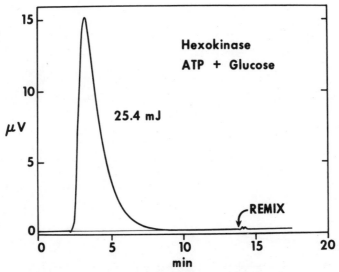

Fig. 8. Chart of calorimeter response for hexokinase catalyzed phosphorylation of glucose in aqueous buffer medium. The small signal labelled remix indicates that reaction was essentially complete on first mixing.

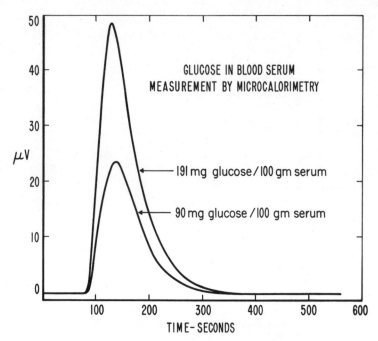

Fig. 9. Chart of calorimeter response for
hexokinase catalyzed phosphorylation of
glucose in human blood serum. Two curves
representing different concentrations of
glucose are shown.

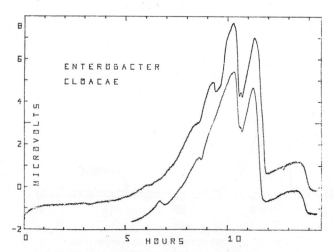

Fig. 10. Repeated growth curves of
Enterobacter cloacae. Curve B is displaced
in time and in response but the scale sizes
remain the same.

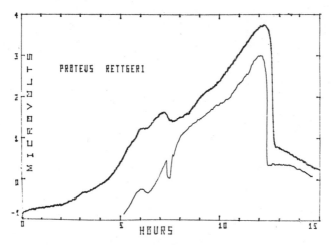

Fig. 11.   Repeated growth curves of Proteus
rettgeri.   The lower curve has been displaced
in time and in response to indicate similarities.   The scale sizes remain the same.

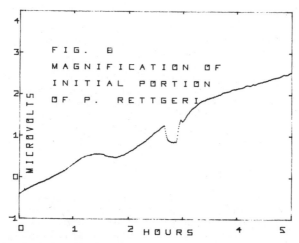

Fig. 12.   Magnification of initial portion of
Proteus rettgeri growth curve.   Note that
rapid changes can be followed by the digital
data acquisition system.

DILATOMETRIC STUDY OF CRYSTALLIZATION AND MELTING OF POLYETHYLENE UNDER HIGH PRESSURE

Y. Maeda and H. Kanetsuna

Research Institute for Polymers and Textiles,
Yokohama, Japan

ABSTRACT

Crystallization and melting behavior of linear polyethylene under high pressures up to 6000 kg/cm^2 has been investigated with a high pressure dilatometer. Crystallization was carried out at a cooling rate of 1°C/min from the melt at each pressure. The samples were characterized by DSC melting behavior, density, and electron microscopy. Folded-chain crystals are formed in the low pressure region below 2000 kg/cm^2. Crystallization in the intermediate pressure region between 2000 and 3500 kg/cm^2 gives a mixture of folded-chain and extended-chain crystals. The extended-chain crystals are the more stable crystal, which predominates with increasing pressure. At high pressures above 4700 kg/cm^2, two stages of crystallization and the corresponding melting of polyethylene can be observed. The phenomenon suggests that the two kinds of extended-chain crystals with different thermal stability, i.e., the ordinary extended-chain crystals and the more extended-chain crystals, form through individual crystallization processes from the melt at high pressure.

INTRODUCTION

In the last decade, many studies have been reported on the thermal properties and the morphology of polyethylene crystallized at high pressures up to 7000 atm. However, it appears that experimental data on the crystallization and melting behavior of polyethylene under pressures above about 2500 atm. already reported by high pressure DTA(1) and dilatometry(2-4) are not consistent with the results from thermal behavior measured at atmospheric pressure(5-9) and the morphology(10-15) of the pressure-crystallized samples. For example, DSC melting curves of the so-called extended-chain crystals formed at pressures above 4000 atm. often show multiple melting peaks in contrast to a single peak in high pressure DTA melting curves (1) or one break in specific volume curves from high pressure dilatometry(2,3). Since it is well known that reduction of pressure from about 30 kilobars to atmospheric pressure at room temperature does not affect the melting behavior and the density (4) of polyethylene, these results are apparently incompatible.

290

More recently, Yasuniwa, et al., reported new data from high pressure micro-DTA(16). At pressures above 4000 kg/cm^2, exothermic and endothermic peaks due to "unknown structure" were observed at higher temperatures than those of crystallization and melting of extended-chain crystals in cooling and heating respectively. Bassett and Turner also reported similar results for fractionated polyethylene(17).

Knowledge of the melting transition of crystalline polymers at high pressure is important because thermodynamic and structural informations can be obtained which experiments at atmospheric pressure do not provide directly. Few studies of polymers have, however, approached the problem from the point of view of crystallization kinetics and the melting porcess, although thermal analysis under high pressure is a useful method of study for crystallization and melting behavior of polymers.

In this study we carried out a dilatometric study of crystallization and melting of linear polyethylene at high pressures up to 6000 kg/cm^2. In addition, density, thermal behavior at atmospheric pressure, and morphology of the samples crystallized under elevated pressure were measured to compare the experimental data so far reported and to understand consistently the pressure-crystallized polyethylene at both atmospheric and elevated pressures.

EXPERIMENTAL

The pressure system used was a modified Koka Flow Tester of Shimadzu Seisakusho Ltd. This apparatus can hold loads up to 80 kg and the force is intensified 10 times large at the plunger head and is fed directly to the pressure vessel capable of sustaining pressures up to 6400 kg/cm^2. The pressure vessel is of the piston-cylinder type equipped with a Bridgman unsupported-area closure as shown in Figure 1. The diameter of the piston is 0.4 cm. Appropriate reduction in cross-sectional area of the piston provides a theoretical pressure magnification of 8. The packing consists of a Teflon ring and a copper ring which backs up the Teflon to give a leak-free movable seal under all conditions used in this study. The temperature is measured by an Alumel-Chromel thermocouple inserted into a hole drilled near the sample cavity in the pressure vessel. Temperature is held within $\pm$ 0.1°C by automatic controll of the electric current input to a resistence ring furnace around the vessel. The temperature measured by the Alumel-Chromel thermocouple was calibrated by using a Copper-Constantan thermocouple. The one junction of the thermocouple was placed in the cylinder and the other was placed in ice and water at 0°C. The electromotive force of the thermocouple was measured to determine the temperature of the sample. At the same time, the Alumel-Chromel thermocouple was used to measure the temperature of the cylinder. The temperature calibration was made by measuring the melting points of pure benzoic acid, indium, and zinc in the cylinder at atmospheric pressure. The temperature difference between the two thermocouples was then measured with polyethylene. In this manner, the actual temperature of the sample was determined to an accuracy of $\pm$ 0.1°C.

Volume changes in specimen were measured by detecting the vertical displacement of the plunger head to a precision + 0.0001 cm with a differential transformer. Because of friction in the packing of the unsupported-area seal, the actual pressure generated in the sample chamber may be different from the pressure calculated using the intensifier ratio. Pressure was calibrated by observing the polymorphic transitions of silver iodide and ammonium iodide which were determined accurately by Bridgman(18). The actual pressure in the sample chamber for a given pressure was determined to an accuracy of ± 50 kg/cm^2. Each dilatometric experiment was repeated to confirm the reproducibility and precision of the melting point and the volume change for fusion of polyethylene. X-ray diffraction patterns of all the pressure-crystallized samples in this study were almost the same as those of the folded-chain crystals formed at atmospheric pressure.

The polyethylene used was Sholex S 6002, a linear polyethylene supplied in pellet form by Showa Denko Ltd. The weight-average and the number-average molecular weights, characterized by gel permeation chromatography, were 165,000 and 31,600, respectively. The experimental procedures were as follows; samples weighing about 70 mg were introduced into the cylinder. After being sealed, the pressure vessel was heated for 30 min at a given temperature under atmospheric pressure to melt the sample completely. The desired pressure was then applied within 3-5 seconds. The pressure was kept constant afterwards during each run. After maintaining the melt at the initial temperature for about 10 min to assure thermal equilibrium, the vessel was cooled at about 1°C/min. It was then heated at a rate of 1°C/min for specific volume measurements. For analysis of the crystal-lized samples, the pressure was released after the temperature was reduced below 30°C and the samples were analyzed under atmospheric pressure.

The density was measured in a density-gradient column of toluene and chlorobenzene at 25°C. The degree of crystallinity based on density was determined by use of the average of Nielson's and Gubler and Kovacs's values(19,20) for the specific volume of the completely amorphous polymer(1.173 cm^3/g) and Swan's value(21) for the specific volume of the ideal crystal (1.001 cm^3/g). The melting of the samples was studied with a Perkin-Elmer DSC 2 differential scanning calorimeter. The calorimeter was run at a rate of 5°C/min with samples weighing 0.3-1.0 mg. The temperature was calibrated to ± 0.5°C by melting point standards(benzoic acid 396.6 K, indium 429.6 K, and lead 600.4 K). The melting temperatures read from the DSC curves were the peak temperatures. Electron microscopy was carried out on standard two-stage replication technique with cellulose acetate as intermediate and chromium shadowed carbon as the final replica. A Jeolco Model JEM 6A transmission electron microscope was used to make observations up to 5000X magnifications.

RESULTS AND DISCUSSION

Figures 2 and 3 show dilatometric curves at pressures up to

3280 kg/cm^2 and between 3500 and 5200 kg/cm^2, respectively. The melting temperature T_m was taken as the point where the sharp increase of specific volume suddenly breaks off to the slowly changing plateau of the molten state. The crystallization temperature T_c was also taken as the point where the linearly decreasing curve of the molten state changes to a sharp decrease by crystallization. This method is relatively sensitive, with an error of less than 1°C. Melting and crystallization temperatures of each crystal except the low-melting solids are summarized in Table I. The dependence of the melting temperature on pressure is shown in Figure 4. The outstanding features of Figures 2-4 are the occurrence of multistage melting phenomena at pressures above 2500 kg/cm^2 and the two-stage crystallization at elevated pressures above 4700 kg/cm^2.

DSC melting point and density of the samples are summarized in Table II. The melting points are not corrected for superheating. Nine DSC curves shown in Figure 5 illustrate the disappearance of the low-temperature peak and the appearance of the intermediate peaks and the high-temperature peak with increasing pressure.

The density increases slightly up to about 2000 kg/cm^2, and between 2000 and 3500 kg/cm^2 it increases rapidly from 0.971 to 0.986 g/cm^3. Further increase in pressure produces a linear increase of density.

Figure 6 is a electron micrograph of a replica of the fracture surface of the sample crystallized at 5130 kg/cm^2. The morphological features are typical of the so-called extended-chain crystals reported previously by many authors. Lamellae with thicknesses up to 1-2 microns have been observed. The thickness of many other lamellae is of the order of 2000-6000 Å.

Figures 2-4 and Tables I and II show clearly that three different kinds of crystals form depending upon temperature and pressure. Following discussion is divided into three parts, covering crystals grown below 2000 kg/cm^2, between 2000 and 3500 kg/cm^2, and above 3500 kg/cm^2. The critical pressure is defined here as the pressure at lower limit observing crystallization under the constant condition of the ordinary extended-chain crystals or the more extended-chain crystals which are named in the paragraph, High Pressure: above 3500 kg/cm^2. The critical pressures were determined by the density and melting point data, and the data obtained separately from the high pressure dilatometry. With respect to the determination of the critical pressure, it is found that density and DSC methods are generally much more sensitive than the high pressure dilatometer used. Our results agree fairly well with those of Wunderlich and Arakawa(5) except for entirely different behavior in the high pressure region above 3500 kg/cm^2.

Low Pressure: below 2000 kg/cm^2

Crystallization and melting in the low pressure region is simple, showing the formation of a crystalline solid. Dilatometric data at pressures up to 2700 kg/cm^2 agree within experimetal error with the results reported by Matsuoka(2) and by Baer and Kardos(3), and also correspond to the results from high pressure micro-DTA reported by Yasuniwa, et al(16). Both density and melting temperature at atmospheric pressure are found to increase slightly for samples crystallized under pressures up to 2000 kg/cm^2. The density increase is rather slight, 0.003 g/cm^3 per 2000 kg/cm^2. Table II shows an increase of about 0.6°C in the melting peak temperature. In all cases, the peak temperature of a single peak was observed at about 134.7 ± 0.5°C, i.e., the melting temperature of folded-chain crystals formed at atmospheric pressure. It appears that the folded-chain crystallization theory(22) is required to explain the crystallization kinetics in the low pressure region. The DSC results agree with the DTA results of Wunderlich and Arakawa (5). As is shown in Figure 4 and Table I, however, the melting data from high pressure dilatometry does not agree with those on extended-chain and folded-chain crystals reported by Davidson and Wunderlich(1). The reason for moderately large discrepancy is not clear at present. The pressure at which the two-stage melting phenomenon can be first observed was 600-800 kg/cm^2 higher than the critical pressure determined from density and DSC. Since the extrapolation to lower pressure of the T_m vs. P curve of folded-chain crystals formed in the intermediate pressure region in Figure 4 intersects the curve of folded-chain crystals in the low pressure region at about 2000 kg/cm^2, the difference in data is attributable to the low precision of the dilatometer. It seems important that the T_m vs. P curve of folded-chain crystals in the low pressure region is continuous with the curve for extended-chain crystals formed in the inter-mediate pressure region.

Intermediate Pressure: 2000 to 3500 kg/cm^2

Dilatometric behavior, density, and DSC curves show marked changes in this region. Although the dilatometric cooling curves as shown in Figures 2 and 3 exhibited an apparently single crystallization process, melting indicated a two-stage melting process. The volume of fusion for the low-temperature melting decreased gradually with increasing pressure. On the other hand, the high-temperature melting was first observed at 2780 kg/cm^2 and the volume of fusion increased progressively with increasing pressure in this region. This behavior has not previously been observed in high pressure dilatometry but has been found in high pressure DTA by Yasuniwa, et al.(16), and by Bassett and Turner(17).

Like the dilatometric melting curves, the DSC melting curves showed two endothermic melting peaks. With the sample crystal-lized at 2200 kg/cm^2, a slight high-temperature peak is observed at about 139°C in addition to the low-temperature peak at 134°C. In contrast to the gradual decrease of the low-temperature peak with increasing pressure, the high-temperature peak developed to the main peak as illustrated in Figure 5. DSC results for samples crystallized in this region also have the same features as the DTA results for isothermally crystallized samples, as reported by Wunderlich and Arakawa(5). According to their interpretation from structural information, the low-temperature and the high-temperature peaks correspond to melting of folded-chain and extended-chain crystals, respectively. Relative area under the high-temperature and the low-temperature peaks determined by DSC and high pressure dilatometry are plotted in

Figure 7 as a function of pressure in the intermediate pressure region. It is clear that melting behavior of folded-chain and extended-chain crystals in the intermediate pressure region shows a good correspondence in measurements between DSC and high pressure dilatometry. The gradual increase in density supports the growth of extended-chain crystals concurrent with the formation of folded-chain crystals. It becomes then necessary to ascertain whether or not the cooling process from the melt can be described as truly single crystallization process. In order to examine qualitatively the crystallization mechanism during slow-cooling more precisely, crystallization from the melt at 3100 kg/cm^2 was stopped at various stages and the crystals formed were heated immediately at 1°C/min until melting went to completion as shown in Figure 8. A volume decrease due to crystallization was clearly observed at about 197°C. For crystallization above 192°C, the subsequent heating curves showed only the melting of extended-chain crystals at 211°C. But for crystallization below 192°C, the heating curves showed a two-stage melting behavior, i.e., melting of folded-chain and extended-chain crystals at 203 and 211°C, respectively. This facts suggest that crystallization on cooling from the melt occurs on two stages, although dilatometric cooling curves show an apparently single curve. Then it must be concluded that at the intermediate pressures folded-chain crystals form subsequently at lower temperature after the crystallization of extended-chain crystals at higher temperature. The experimental facts seem to support the suggestions reported by Bassett, et al.(23), that there are independent processes for folded-chain and extended-chain crystallizations.

High Pressure: above 3500 kg/cm^2

Crystallization and melting at pressures up to 4700 kg/cm^2 were found to proceed much as in the upper intermediate pressure region. The dilatometric heating curves show a small shoulder on the main fusion curve. At high pressures above 4700 kg/cm^2, it is found, surprisingly, that crystallization from the melt takes place through a two-stage process, and that melting occurs by a corresponding two-stage process together with one or two low-temperature shoulders. The volume decrease at the first-stage of crystallization is small, while that at the second-stage of crystallization is always large. The volume ratio of fusion of the high-melting and the low-melting portions was about between 1:9 and 2:8, independent of pressure. The melting temperature of the new high-melting portion under high pressure agrees with the melting data for the so-called extended-chain crystals, as reported by several authors(1,3,24). On the other hand, the T_m vs. P curve of the extended-chain crystals formed in the intermediate pressure region also connects smoothly with the curve for the new high-melting portion in the high pressure region. The extrapolation to lower pressure of the T_m vs. P curve for the low-melting portion of the ordinary extended-chain crystals formed in the high pressure region intersects the curve for the extended-chain crystals formed in the intermediate pressure region at about 3500 kg/cm^2. The results support the concept that the critical pressure of the new high-melting portion exists at about 3500 kg/cm^2.

The morphological observations shown in Figure 6 indicate that extended-chain lamellae several thousand angstroms thick predominate but crystals more than one microns thick are scarce.

In contrast to the occurrence of the peculiar phenomenon in high pressure dilatometry, however, no DSC peak is observed above the high-temperature peak of the so-called extended-chain crystals. The melting behavior of the samples crystallized above 3500 kg/cm^2 is complicated; one or two peaks are observed at 136-8°C between the low-temperature and the high-temperature peaks. The high-temperature peak appears at 141.5°C regardless of pressure above 4000 kg/cm^2.

Similar behavior mentioned above has been described briefly in recent articles by Yasuniwa, et al.(16), and by Bassett and Turner(17). They considered the first-stage of crystallization at pressures above 4700 kg/cm^2 as a first order transition between the melt and a kind of mesophase which appears at elevated pressures above 4000 kg/cm^2. These propositions of the existence of mesophase under high pressure have been made in the circumstances of understandings on extended-chain crystals of which the thermal behavior at atmospheric pressure and morphology could not give unambiguously informations on the peculiar phenomenon. However, the following experiment, which will be discussed in detail elswhere, will supplement the morphological observations. Isothermally crystallized sample formed at 235°C and 5150 kg/cm^2 for 20 hr were treated with fuming nitric acid at 80°C for 50 hr. The analysis of the GPC curve of the treated sample identified a small peak as polymers with average molecular chain length of about 2 microns, but the main peak corresponded to a molecular length of about 3000 Å, which corresponds to the average lamellar thickness of the so-called extended-chain crystals observed by elctron microscopy(10,14,15). The order of the average extended-chain length for the small peak is consistent with morphological observations of fracture surfaces. Thus, the occurrence of the small peak in the corrected GPC curve indicates that another kind of extended-chain crystals exist in the pressure-crystallized sample. Although the DSC results show no evidence of the another extended-chain crystals, the main peak observed at 141-2°C is believed to represent the overall melting of both the extended-chain crystals. The apparent two-stage crystallization and the corresponding melting behavior indicate that two kinds of extended-chain crystals have grown through individual crystallization processes respectively. Therefore, the new high-melting portion formed through the first-stage of crystallization with part of a sample may be defined as the more extended-chain crystals that are distinguished from the ordinary extended-chain crystals formed through the second-stage of crystallization. From this point of view, thermal behavior at atmospheric pressure, density, and morphology of pressure-crystallized samples can be explained in a way consistent with the crystallization and melting behavior of polyethylene under high pressure. In any event, the occurrence of these crystallization and melting phenomena at pressures above about 4000 kg/cm^2 will significantly affect arguments concerning the two theoretical explanations(8,12,15,25) of crystallization mechanism leading to the so-called extended-chain crystals.

To conclude, folded-chain crystals are formed as a stable
solid at pressures up to 2000 kg/cm^2. As depicted in Figure 5,
extended-chain crystals are formed at high pressures above 2200
kg/cm^2 and develop extensively with increasing pressure in
contrast to the decrease of folded-chain crystals. Mixed crystallization of extended-chain and folded-chain crystals takes place
in the intermediate pressure region. At higher pressures above
4700 kg/cm^2, a new phase can be observed as illustrated in
Figure 3. It is most probable that the new phase is not a kind
of mesophase but the most stable crystal, i.e., the more extended-
chain crystal, existing in the crystals formed under elevated
pressures up to 6000 kg/cm^2.

REFERENCES

1.   T. Davidson and B. Wunderlich, J. Polym. Sci. A-2, 7, 377(1969)
2.   S. Matsuoka, J. Polym. Sci., 57, 569(1962)
3.   E. Baer and J. L. Kardos, J. Polym. Sci. A, 3, 2827(1965)
4.   J. Osugi and K. Hara, Rev. Phys. Chem. Japan, 36, 28(1966)
5.   B. Wunderlich and T. Arakawa, J. Polym. Sci. A, 2, 3697(1964)
6.   J. L. Kardos, E. Baer, P. H. Geil and J. L. Koenig,
     Kolloid-Z., 204, 1(1965)
7.   R. B. Prime, B. Wunderlich and L. Melillo, J. Polym. Sci.
     A-2, 7, 2091(1969)
8.   P. D. Calvert and D. R. Uhlmann, J. Polym. Sci. B, 8, 165(1970)
9.   H. Kanetsuna, S. Mitsuhashi, T. Hatakeyama, M. Iguchi,
     M. Kyotani and Y. Maeda, The IUPAC Symposium on Macromolecules
     in Helsinki, Finland(July, 1972).
10. P. H. Geil, F. R. Anderson, B. Wunderlich and T. Arakawa,
     J. Polym. Sci. A, 2, 3707(1964).
11. B. Wunderlich and L. Melillo, Makromol. Chem., 118, 250(1968)
12. B. Wunderlich and T. Davidson, J. Polym. Sci. A-2, 7, 2043(1969)
13. T. Davidson and B. Wunderlich, J. Polym. Sci. A-2, 7, 2051(1969)
14. R. B. Prime and B. Wunderlich, J. Polym. Sci. A-2, 7, 2061(1969)
15. D. V. Rees and D. C. Bassett, J. Polym. Sci. A-2, 9, 385(1971)
16. M. Yasuniwa, C. Nakafuku and T. Takemura, Polymer J., 4,
     526(1973)
17. D. C. Bassett and B. Turner, Nature Physical Science, 240,
     146(1972)
18. P. W. Bridgman, Proc. Amer. Acad. Arts Sci., 51, 55(1915);
     idem., ibid., 52, 91(1916)
19. L.E. Nielson, J. Appl. Phys., 25, 1209(1954)
20. M. G. Gubler and A. J. Kovacs, J. Polym. Sci., 35, 551(1959)
21. P. R. Swan, J. Polym. Sci., 56, 403(1962)
22. J. D. Hoffman, J. L. Lauritzen, Jr., F. Passaglia, G. S. Ross,
     J. F. Frolen and J. J. Weeks, Kolloid-Z., 231, 564(1969)
23. D. C. Bassett, B. A. Khalifa and B. Turner, Nature Physical
     Science, 239, 106(1972)
24. D. V. Rees and D. C. Bassett, J. Polym. Sci. B, 7, 273(1969)
25. J. L. Kardos, H. M. Li and K. A. Huckshold, J. Polym. Sci.
     A-2, 9, 2061(1971)

TABLE 1

Melting and crystallization of polyethylene under
high pressure

Pressure kg/cm^2	Folded-chain crystals		Ordinary extended-chain crystals		More extended-chain crystals	
	T_m,°C	T_c,°C	T_m,°C	T_c, °C	T_m,°C	T_c,°C
630	161	149.5				
1060	173	156				
1490	181.4	166.5				
1950	190.3	174				
2480			202.1	183		
3050	200.2		209.3	189.5		
3500	212.9		217.6	199.3		
3910	218.3				226.1	206
4530	221.9				238.5	221
4930			236.7	224.3	243.6	228.2
5570			249	231.5	255.6	238
6380			257.2	239.5	269.8	250

TABLE 2

Data of polyethylene crystallized under various pressures

Crystallization pressure kg/cm^2	Density d(25°C) g/cm^3	Cryst., %	DSC results			
			Low temp. peak,°C	°C		High temp. peak,°C
220	0.9680	81.4	134.3			
630	0.9685	81.7	134.7			
1060	0.9688	81.9	134.5			
1490	0.9702	82.9	134.9			
1950	0.9706	83.0	135.2			
2200	0.9709	83.2	134.9			139.6
2480	0.9724	84.1	134.9			139.9
3050	0.9768	86.8	134.7			139.5
3280	0.9828	90.4	133.7			139.9
3500	0.9860	92.3	133.0			140.4
3710	0.9873	93.1	131.7			140.3
3910	0.9868	92.8	131.5	135.4	137.6	140.6
4130	0.9876	93.3	131.4	136.0	138.1	141.2
4330	0.9886	93.9	131.7	135.6	138.2	141.2
4740	0.9906	95.1	129.9	135.7		141.4
5130	0.9922	96.0	129.8	135.5		141.5

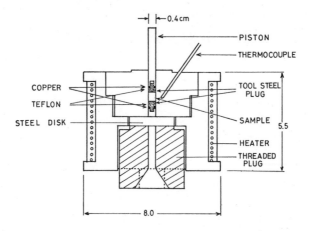

Fig.1   High pressure vessel and seal assembly.

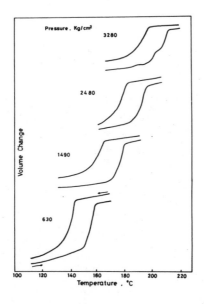

Fig.2 Volume-temperature curves at various high pressures
ranging from 630 to 3280 kg/cm^2 . The samples were crystallized from the melt with a cooling rate of 1°C/min at each
pressure. Melting experiments were subsequently performed
at 1°C/min at the same pressures.

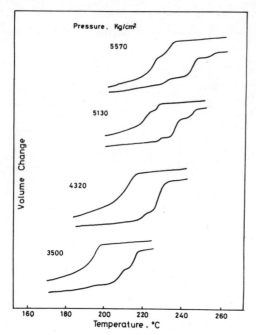

Fig.3 Volume-temperature curves at various high pressures ranging from 3500 to 5570 kg/cm^2. Cooling and heating processes were performed by the same procedures as in Fig.2.

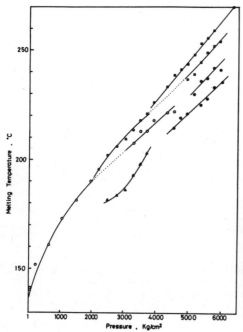

Fig.4 Melting temperature vs. pressure: (O) folded-chain crystals, (□) extended-chain crystals, (■) more extended-chain crystals, (△) low-melting crystals formed in the intermediate pressure region, (◑) and (◐) low-melting crystals formed in the high pressure region.

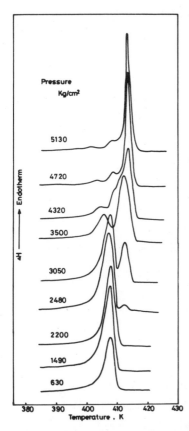

Fig.5 DSC melting curves
of polyethylene crystal-
lized at indicated pres-
sures. DSC curves meas-
ured at 5 °C/min under atmospheric pressure were
not corrected for instru-
mental lag.

Fig.6 Electron micrograph
of a replica of fracture
surface of polyethylene
crystallized slowly from
the melt at 5130 kg/cm^2.

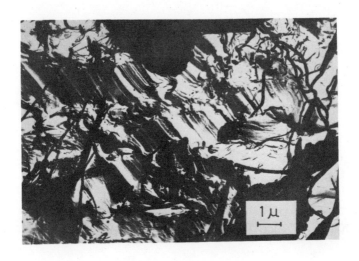

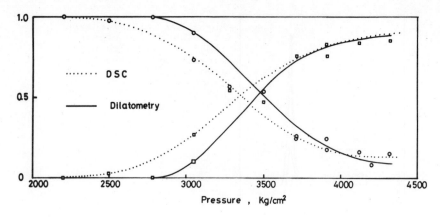

Fig.7 Effect of pressure on the amount ratio of melting of the folded-chain and the extended-chain crystals in the intermediate pressure region, investigated separately by DSC and high pressure dilatometry. Symbols used are identical with in Fig.4.

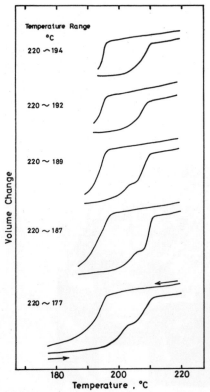

Fig.8 Volume-temperature curves of polyethylene at 3100 kg/cm². The samples were crystallized from the melt slowly in the indicated temperature ranges and followed by heating at 1 °C/min to melt the sample.

THE APPLICATION OF THERMAL ANALYSIS
IN THE AEROSPACE INDUSTRIES

Henry L. Friedman

General Electric Company, Re-Entry and Environmental
Systems Division, P.O. Box 8555, Philadelphia, Pa.   19101

ABSTRACT

Several different thermal analysis measurements are used at General
Electric in the design and flight evaluation of re-entry vehicle heat shield charring
ablative polymers. The various properties are discussed together with laboratory
apparatus and techniques that are used to measure them. The practical limits of
uncertainty permitted for some of the parameters are also discussed. Several
other aerospace applications of thermal analysis methods as used by other compan-
ies are described, including the surface analyzer for the 1976 Mars Viking Lander,
the analysis of lunar materials and the analysis of toxic compounds from fires which
could occur in manned satellites and spacecraft.

INTRODUCTION

The aerospace industry started in the mid 1950's, shortly after the start of
the availability of modern commercial thermal analysis equipment. Since the first
major problem area was re-entry heating, it is not surprising that thermal analysis
has played a very major role in the development of aerospace technology. This
paper is divided into two main parts. The first is a discussion of ablation science,
with special emphasis on char forming polymers. Theoretical aspects of thermal
material behavior will be described, and some examples will be given of the application of thermal methods to provide data necessary for design and performance
evaluation purposes. The second part presents descriptions of special utilization
of thermal methods to other problem areas that are quite unique to the aerospace
industries.

ABLATION SCIENCE

While this discussion emphasizes the ablation of char forming polymers as
re-entry vehicle heat shields, many elements are of major importance in other
related applications of materials, e.g., ablative rocket nozzle liners, combustion
of solid rocket propellants and other combustible solids.

303

Consider a body of sphere-cone configuration entering into the Earth's atmosphere from outer space at a very high velocity. Because of air friction and in some cases radiation from the shock layer, the ablator begins to heat to very high temperatures, which triggers an extremely complicated set of events. As the material heats, it begins to decompose, melt or sublime. For char forming plastics, decomposition includes gasification and progressive modification of the solid until the residue is completely carbonized. Carbon, of course, is an extremely refractory material. The interior carbonizes, progressively. As the decomposition gases flow through the solid, they react with each other and with the surfaces. The mechanical properties of the solid vary continuously. Eventually, the gases flow out of the solid and merge with the stream of gases from the atmosphere. The decomposition gases react with the environment, and so do the environment gases interact with the surface of the heat shield.

Mathematical analysis of these phenomena are very complicated, and involve many interactions. Some of the earliest and most significant studies in ablation science were carried out at General Electric by Dr. S. M. Scala and his coworkers (l). A view of the state-of-the-art may be gained from reference 2. The gross physical and chemical phenomena and their interactions are outlined in Table 1. Clearly, in order to understand the intricacies of the phenomena involved, one would have to understand the details of the various processes. The thermal methods have supplied much of these data. Table 2 presents a summary of various properties that have been measured by thermal analysis, and relates them to factors that are important in re-entry heating.

As a practical matter, the interactions of the various phenomena as well as the uncertainties are such that it is not necessary to employ the ultimate in sophistication and accuracy in obtaining or analyzing the laboratory thermal data. Shaw et al. (3) have demonstrated the effect on ablative thickness required, of uncertainties in thermophysical properties for several different configurations flying a variety of missions. The following discussion describes thermal analysis apparatus available in our laboratories and some aspects of their specific use and data analysis.

Wherever ablation calculation results are presented in this paper, they were obtained with General Electric's Reaction Kinetics Ablation Program (REKAP), a semi-empirical digital computer analysis program that simulates the decomposition processes associated with charring ablators. It includes the three significant regions of interest in the interaction and coupling between the material degradation and hypersonic environment. These include the gas-phase boundary layer, the condensed phase and the interaction between phases. Figure 1 is a schematic of the regions involved, and representative temperature, density and pressure variations across the char layer and condensed phase. The equations and assumptions that were used may be found in reference 3.

Thermogravimetry

Our main TG measurements have until recently been performed with a Fisher Thermobalance (based on a Cahn Model RG Electrobalance) with conventional programming capabilities of from 0.5 to 30° C/min. to about 1000° C, and with heating rates up to 90° C/min. achieved by a motor driven autotransformer. The fast heating rates are used to simulate the very rapid heating during re-entry. The laboratory recently acquired a Rigaku-Denki Thermoflex TG/DTA unit for simultaneous measurements of these parameters. Heating rates ranging from 0.625 to 160° C/min. are available, with a maximum temperature of 1500° C.

Since re-entry heating is sometimes much more rapid than that available in conventional thermobalances, Melnick and Nolan (4) developed a thermobalance capable of heating rates up to 100° C/sec.

The balance portion of the apparatus consisted of four active strain gauges arranged in a Wheatstone bridge configuration. The gauges were cemented to a strip of 10-mil thick stainless-steel shim stock from which the sample hung. Two gauges on top measured the contraction forces, and the two located on the bottom measured the expansion of the shim stock as the sample was heated. The balance was designed to be independent of sample sizes. Weights up to 10 g could be accommodated with a detectable weight loss of $\pm$ 1 mg.

Since the bottom strain gauges were subject to thermal gradients during a heating cycle, a 2000-Ω thermistor was made part of a separate Wheatstone bridge to obtain a signal proportional to the temperature of the gauges. A fraction of this signal was selected by a pair of potentiometers and was fed into a stable dc microvolt amplifier. The amplifier output was electronically subtracted from the strain gauge weight-loss signal which was read out on the Y axis of a high-speed X-Y recorder or an oscillograph.

The source of thermal energy was a vertically mounted infrared radiant heater with elliptical reflectors. Lamps used in this heater peaked at a wavelength of 5000-5800 Å. Heating rates were changed by transformer control off the 440-V ac line. Heat fluxes at heating rates of 100° C/sec have been shown to be in excess of 17 BTU/ft^2-sec as measured by a copper calorimeter placed at the location of the sample in the system.

Results for a nylon-reinforced phenolic sample at 10° C/min. and 100° C/sec. are shown in Figures 2 and 3, respectively. Note the drastic differences in the shapes of the thermograms.

Kinetic treatment of TG data is based on the following form of the Arrhenius equation

$$-(1/w_0)\,(dw/dt) = Ae^{-\Delta E/RT} \left[(w-w_f)/w_0 \right]^n$$

where w = weight, w_0 = initial weight, w_f = final weight, t = time, T = absolute temperature, R = gas constant, A = pre-exponential factor, ΔE = activation energy and n = order of reaction. TG is normally obtained over a wide range of heating rates. Kinetic analysis of a set obtained for a phenolic in the range 4.5 to 62° F/min. are shown in Figure 4. Integer orders were assumed, and the values that produced the most consistent set of rate constants ($k = Ae^{-\Delta E/RT}$) were accepted as being appropriate. The best fit for the data in Figure 3, was obtained for two separate reactions with n = 2 in both cases.

General Electric scientists have made numerous contributions to kinetic analysis of thermal analysis data (5-11).

The total fractional mass of polymer converted to char may be determined from the final TG weight under proper experimental conditions. In the event no appreciable swelling or shrinkage occurs during pyrolysis, the TG data also provides a continuous evaluation of the change of density. The relationship between char density and kinetics of polymer degradation will be discussed in the section on char density.

Evolved Gas Analysis

The compositions of gases evolved from degrading polymers have been measured by a variety of techniques at General Electric. In our earliest research, samples of ablative materials were heated in vacuo by an arc-image furnace, and gases were fractionated and were analyzed by gas chromatography, mass spectrometry and other methods (12, 13). Later, polymer samples were subjected to flash pyrolysis in the reaction chamber of a Bendix Time-of-Flight Mass Spectrometer with rapid analysis following the flash (14). Then, ablative polymers were heated resistively in a notch in a graphite filament located in a differentially pumped zone near the electron beam of a Bendix T-O-F Mass Spectrometer. Mass spectra were continuously scanned every 0.1 sec. and were recorded with an oscillograph. Heating rates of up to 1000° C/sec. were achieved by this method (15). A typical set of results for a phenol-formaldehyde polymer is shown in Figures 5 and 6, and were ascribed to H_2(m/e 2), N_2 (m/e 14 and 28), NH_3(m/e 14-17), CH_4 (m/e 14-16), H_2O(m/e 16-18), HCN(m/e 27), CO(m/e 28), C_2H_4(m/e 27 and 28), CO_2(m/e 16, 28 and 44), phenol(m/e 94), cresols (m/e 108), and xylenols (m/e 122).

Laser (16) and inductive (17) heating have also been used with continuous T-O-F mass spectrometry to study species evolved from the vaporization of carbon and other ablative materials. Inductive heating with gas chromatography has been used to study the transpiration chemistry of primary polymer degradation products passing through porous chars.

The most sophisticated EGA-MS measurements carried out at General Electric have employed linear programmed rates of 10° C/min. in a quartz tube furnace followed by continuous gas analysis with a Bendix Time-of-Flight Mass Spectrometer (15, 18-21). Two automatic data processing systems have been used with these measurements. The first used a stepped programmable mass selector with all data recorded on perforated paper tape after processing through an analog-to-digital converter. Subsequent data processing used several digital computers and an automatic graph plotter (15, 18-20). A typical page of data for an aromatic polyamide is shown in Figure 7. It was customary to scan some 200 mass peak locations with that system. The later data recording was on FM analog magnetic tape with subsequent data processing with a hybrid computer. Mass peaks were read automatically and were assigned mass numbers by the computer (21). A typical EGA-MS curve for m/e-27 (HCN) for a polyquinaxaline is shown in Figure 8. This system was used to provide compositions of decomposition gases evolved from ablative polymers, and was valuable in explaining sources of un-expected emission spectra observed from instrumented re-entry vehicle heat shields.

If one were to require extremely fast rates of temperature rise, the system used by Stapleton (22) at Lockheed Propulsion Corp. should be considered. Solids deposited on a platinum filament (which was a leg of a Wheatstone bridge) were heated at rates up to 10^6 ^{o}C/sec. by capacitor discharges, with product analysis by a Bendix T-O-F Mass Spectrometer.

So far the accuracy of kinetics obtained from TG data has been adequate for ablation modeling, (3) even for phenolics, where the actual decomposition is known to involve many more than two stages, as observed in Figure 4. If and when the computational tools become sensitive enough to require better resolution, EGA-MS will have much to offer towards comprehensive kinetic analysis. Take, for example, the EGA-MS measurements for a phenolic that were observed by Shulman and Lochte (23) and are shown in Figure 9. Note the different tempera-ture ranges and shapes of the evolution curves for the 11 products that were ob-served by those workers. Arrhenius plots for their kinetic analysis are shown in Figure 10, which are based on the assumption that at the start of evolution,

Δ E could be evaluated independent of n and A. Of course, one needs a more sophisticated analysis to obtain all the kinetic parameters, but all of the data that are needed are available. It would be most desirable to carry out runs over a wide range of heating rates in order to observe changes that could result there-from, especially if there are competitive reaction mechanisms.

For ablation calculations, EGA is important for estimating the average molecular weight of the species injected into the boundary layer. Figure 11 shows the effect of gas molecular weight on the weight of a nylon reinforced phenolic heat shield for a satellite which re-enters the Earth's atmosphere in a ballistic trajectory. The altitude vs. velocity profile for this trajectory is included in Figure 12 and aerodynamic environments are shown in Figure 13. In examining the data of Figure 11, it is important to realize that the weight of a satellite or missile system is of extreme importance relative to the rocket power required and for the weight of the other components that may be carried.

Differential Scanning Calorimetry

The Perkin-Elmer model DSC-1 differential scanning calorimeter is commonly used to determine specific heats, heats of transformation, heat of decomposition and for quality control of ablative materials. These thermophysical properties are of extreme importance in ablation calculations. Figure 14 shows the effect of environment on the enthalpy of gasification of ESM 1030 (one of General Electric's siloxane elastometer based ablative materials). These data were obtained by combining DSC and TG data obtained at the same atmosphere and in the same environments. The heat transfer is clearly extremely dependent on the atmosphere, so this knowledge is vital for accurate ablation calculations.

General Electric scientists have been active in quantitative DTA, an area closely related to DSC (24-26).

Char Density

The aerothermodynamicist considers char to cover the entire range from very minutely decomposed polymer to fully carbonized residue, whether these be porous or non-porous. Therefore, the char yield as obtained from TG data has only limited value in ablation calculations (3). Much more useful are post test examinations of bulk density of solid residue of materials that have undergone ablation under simulated re-entry environments, as for example in Figure 15. The density profile, for the refactory reinforced phenolic ablator which was tested under conditions shown in the figure, was obtained by a series of high-speed, fine grit grinding operations, with successive cleaning and weighing operations. Note that the two stage kinetic treatment (Figure 5) provides a much better data fit to actual observations than the best single kinetic equation fit obtained by Shaw et al. (3). Those authors attributed the decrease in char density near the surface to oxidation in depth, during and after the test. The present author feels that sublimation of carbon could have been a further cause of the discrepancy.

Thermal Conductivity and Emittance

The various instruments that are used to determine thermal conductivity and emittance of solids are described in reference 27, together with some specific applications. Shaw and coworkers (3) demonstrated that uncertainties in thermal conductivity of partly decomposed material and/or char had a major effect on the required weight of an ablative heat shield.

308

OTHER AEROSPACE APPLICATIONS

Planetary Surface Analysis

The 1976 Mars Viking Lander will carry a pyrolysis-gas chromatography-mass spectrometry system which is designed to analyze organic compounds, water and volatile constituents in the atmosphere and surface of the planet (28). The instrumental aspects and data interpretation of the system associated with organic compounds in the surface are based on the efforts of Shulman, Simmonds and coworkers (29-32) at Jet Propulsion Laboratory. A functional block diagram of the system is shown in Figure 16. Volatile compounds will be identified intact, while involatile ones will be inferred from the character of their pyrolysis products. The system is very sensitive, but has safeguards against overloading (effluent divider). Hydrogen is the GC carrier gas, which is removed through a heated thin-walled palladium-silver alloy tube just before the mass spectrometer. The removal process will cause some changes in the gases analyzed by MS. Soil samples will be heated in the carrier gas stream at 150, 300 and 500° C in tiny cylindrical furnaces. Direct EGA-MS will be used for compounds that can be volatilized out of the soil sample, but that would irreversibly condense in the GC column.

A separate DSC-EGA system (Figure 17) was originally proposed by the Bendix Corp. for determination of the forms of occurrence of H_2O in the Martian surface materials (33). While the technical feasibility of the approach was demonstrated, as shown in Figure 18, funding limitations necessitated its omission from the project. Therefore, H_2O will be analyzed by the GC-MS system, with the limitation that sample transport and processing will occur at temperatures higher than Martian ambient, so loosely bound water, if present, will almost surely be lost. The step heating of the pyrolysis-GC-MS system will also limit the data interpretation capabilities compared to programmed heating for DSC.

The mass spectrometer will be the Nier-Johnson double focusing instrument with mass range of 12-200. The laboratory instrument tested in 1971 had a sensitivity of 2.5×10^{-8} amp sec/ng at a multiplier gain of 10^4. The GC consists of two columns in series, the first packed with 10% SF 96 on Chromosorb W followed by a stainless steel capillary coated with 10:1 SF 96: Igapal CO 880. The overload protecting device is actuated by the total ion monitor and the ion amp current, and consists of a stream splitter which can vent more than 99.9% of the GC effluent to the Martian atmosphere. The GC detector is of the cross-section type to permit any molecule to be detected. It is less sensitive than the MS detector.

Since local heating in the landing exhaust plume impingement area may affect the composition of the material to be tested, research was carried out on site alteration, which results suggested certain changes in fuel composition. The authors felt that by intercomparing the data gathered over the full 90 day mission,

309

they would be able to distinguish between indigenous constituents and those created by the lander during descent. Very elaborate precautions are being taken to minimize contamination from various potential sources.

MTA for Investigation of Various Lunar Materials

Gibson and coworkers (34-37) at the National Aeronautics and Space Administration's Manned Spacecraft Center have used a system based on the Mettler vacuum recording thermoanalyzer, the Finnigan 1015 S/L quadrupole mass spectrometer, a modified Systems Industries System/150 gas chromatograph - mass spectrometer computer system, and a computer-mass spectrometer interface to study various lunar materials. The results shown in Figure 19 were for an Apollo 15 soil which was stored in a nitrogen atmosphere until MTA was performed in vacuo. These tests permitted the authors to observe gas evolution from (1) atmospheric contaminants, (2) chemical reactions between solid phases, (3) solar wind bombardment, and (4) gases from vesicles and inclusions and/or gases exsolving from the melt. They also irradiated terrestrial soils with 100 KEV protons to simulate a solar wind and were able to observe H_2O evolution that resulted therefrom. The MTA water evolution pattern was similar to those observed from some lunar samples.

Toxic Compounds in Manned Satellite and Aircraft Cabin Fires

Kleinenberg and Geiger (38, 39) at the Aerospace Medical Research Laboratory have employed TG-EGA-MS in order to study gases evolved from polymeric material during fires, for the purpose of toxicological evaluation of their impact on the environment in satellites and aircraft. The unique aspect of their work is that they sample evolved gases while the polymer sample is heated in an oxidizing atmosphere at ambient pressure. This is accomplished by use of a long stainless steel capillary tube whose entrance is about 5 mm above the sample. The capillary opening is constricted so that the gas pressure inside the ion source of a Bendix T-O-F Mass Spectrometer is 2 to 3 x 10^{-5} torr when the tip is at atmospheric pressure. The dimensions are such that viscous flow results inside the capillary, which prevents fractionation. To prevent condensation inside the capillary, the line is heated resistively with a DC power supply.

In their earliest system these authors used a Cahn Model RH Electrobalance as the basis of the thermobalance, with the T-O-F mass spectrometer (38). Recently they replaced the mass spectrometer with a duPont model 21-491B Mass Spectrometer and the model 21-094 Mass Spectrometer Data System with library search capability (39). The authors found the data system to be extremely valuable for their work.

The present author believes that EGA-MS in the way that is used here is extremely useful for studying gases released during fires. However, it should be pointed out that it is not possible to simulate the complexities of real fires with such a simple experiment. The chemical compositions and relative quantities of products will depend on whether they are evolved from pyrolysis or surface oxidation, and the specific natures of the environments where gas combustion occurs. Aerothermochemical analysis would be helpful, but even in that case geometric simplification would be necessary. Lew (40, 41) has studied the ignition and sustained combustion of plastics in gravity and reduced gravitational fields in typical space cabin atmospheres. Lew's computational model couples material and gas interactions.

ACKNOWLEDGEMENTS

The author is grateful to Mr. D. Florence for his helpful discussions on the present state-of-the-art of the application of thermal analysis data for ablation modeling. He is also indebted to Mr. K. W. Bleiler and to Mr. J. P. Brazel for preparing materials on apparatus available for routine thermal analysis at G. E. Re-entry and Environmental Systems Division.

Figures 2 and 3 were reprinted from reference 4, page 647, by courtesy of Marcel Dekker, Inc. Figure 7 was reprinted from reference 19, p. 415 by permission of Academic Press. Figure 8 was reprinted from reference 21, page 125 by permission of Elsevier Scientific Publishing Co. Figures 9 and 10 were reprinted from reference 23, pages 621 and 622, respectively, by permission of John Wiley and Sons, Inc. Figures 16, 17, and 18 were reprinted from reference 28, pages 117, 127, and 128, respectively, by permission of Academic Press and the authors. Figure 19 was reprinted from reference 36, page 2035, by permission of MIT Press.

REFERENCES

1.　　S. M. Scala, in Proc. 10th Int. Astronautical Congress, London, England, Sept. 1959

2.　　W. H. Dorrance, Viscous Hypersonic Flow, Mc-Graw-Hill, N. Y., 1962

3.　　T. E. Shaw, D. Garner and D. Florence, AIAA Paper No. 65-639, Presented at AIAA Thermophysics Specialist Conf., Monterey, Calif., Sept. 13-15, 1965

4.　　A. M. Melnick and E. J. Nolan, J. Macromol. Sci.-Chem., A3, 641 (1969)

5.　　C. D. Doyle, J. Appl. Polymer Sci., 5, 285 (1961)

311

6. Ibid., 6, 639 (1962)

7. H. L. Friedman, J. Polymer Sci., C6, 183 (1964)

8. C. D. Doyle, Makromol. Chem., 80, 220 (1964)

9. C. D. Doyle, Nature, 207, 290 (1965)

10. C. D. Doyle, Tech. Methods Polym. Eval., 1, 113 (1966)

11. H. L. Friedman, Polymer Letters, 7, 41 (1969)

12. H. L. Friedman, in Re-entry and Vehicle Design, D. P. LeGalley, ed.,
    Academic Press, N. Y., 1960, Vol. 4, p. 3

13. H. L. Friedman, J. Appl. Polymer Sci., 9, 1005 (1965)

14. Ibid., 651

15. H. L. Friedman, H. W. Goldstein and G. A. Griffith, presented at
    the 15th Ann. Conf. on Mass Spectrometry and Allied Topics, Denver,
    Colo., May 1967 (Proceedings p. 16)

16. P. D. Zavitsanos, Carbon, 6, 731 (1968)

17. P. D. Zavitsanos and G. A. Carlson, J. Chem. Phys., 59, 2966 (1973)

18. H. L. Friedman and G. A. Griffith, in Thermal Analysis 1965,
    J. P. Redfern, ed., Macmillan, London, 1965, p. 22

19. H. L. Friedman, G. A. Griffith and H. W. Goldstein, in Thermal
    Analysis, P. D. Garn and R. F. Schwenker, Jr., eds., Academic Press,
    N. Y., 1969, Vol. 1, p. 405

20. H. L. Friedman, Thermochim. Acta, 1, 199 (1970)

21. H. L. Friedman, G. A. Griffith, J. R. Mallin and N. M. Jaffe,
    ibid., 8, 119 (1974)

22. W. G. Stapleton, presented at the 17th Ann. Conf. on Mass Spectrometry
    and Allied Topics, Dallas, Texas, May 1969 (Proceedings p. 128)

23. G. P. Shulman and H. W. Lochte, J. Appl. Polymer Sci., 10, 619 (1966)

24. D. M. Speros and R. L. Woodhouse, J. Phys. Chem., 67, 2164 (1963)

25.   Ibid., 72, 2846 (1968)

26.   D. M. Speros, in Thermal Analysis, P. D. Garn and R. F. Schwenker, Jr., eds., Academic Press, N. Y., 1969, Vol. 2, p. 1191.

27.   Thermophysical Characterization, General Electric Rept. No. TIS 74SD231, to be published.

28.   D. M. Anderson, K. Biemann, L. E. Orgel, T. Owen, G. P. Shulman, P. Toulmin, III and H. C. Urey, Icarus, 16, 111 (1972)

29.   G. P. Shulman and P. G. Simmonds, Chem. Comm., 1968, 1040

30.   P. G. Simmonds, G. P. Shulman, and C. Stenbridge, J. Gas Chromatog., 7, 36 (1969)

31.   P. G. Simmonds, Appl. Microbiol., 20, 567 (1970)

32.   P. G. Simmonds, G. R. Shoemake and J. E. Lovelock, Anal. Chem. 42, 881 (1970)

33.   Viking Soil Water Feasibility Study, Bendix Sci. Rept. 2795, Bendix Aerospace Syst. Div., Ann Arbor, Mich., 1969

34.   E. K. Gibson, Jr. and S. M. Johnson, in Proc. 2nd Lunar Sci. Conf., A. A. Levinson, ed., MIT, Cambridge, Mass., 1971, Vol. 2, p. 1351

35.   E. K. Gibson, Jr. and S. M. Johnson, Thermochim. Acta, 4, 49 (1972)

36.   E. K. Gibson, Jr. and G. W. Moore, in Proc. 3rd Lunar Sci. Conf., D. Heymann, ed., MIT, Cambridge, Mass., 1972, Vol. 2, p. 2029

37.   E. K. Gibson, Jr. and G. W. Moore, Science, 179, 69 (1973)

38.   G. A. Kleinenberg and D. L. Geiger, in Thermal Analysis, Proc. 3rd Int. Conf. Thermal Anal., H. G. Wiedemann, ed., Birkhauser Verlag, Basel, 1972, Vol. 1, p. 325

39.   D. L. Geiger and G. A. Kleinenberg, presented at the 21st Ann. Conf. on Mass Spectrometry and Allied Topics, San Francisco, Calif., May 20-25, 1973. (Proceedings p. 182)

40.   H. G. Lew, Investigation of the Ignition and Burning of Materials in Space Cabin Atmospheres. I. Ignition and Burning of Material, Rept. No. NASA-CR-128068, Mar. 1972

41.   Ibid, II. Ignition of a Combustible Mixture by a Hot Body with the Effect of Gravity, Rept. No. NASA-CR-128064, Mar. 1972

TABLE 1

Aerothermochemistry of Materials in Hypersonic Flows

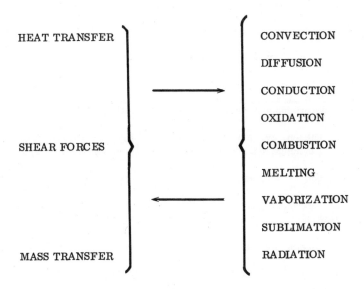

TABLE 2

PROPERTY	SIGNIFICANCE	HOW DETERMINED
1. Kinetics of thermal decomposition of ablative solid.	Rate of vaporization; gas input term for convection, diffusion, conduction and radiation; fuel input for combustion.	TG or EGA
2. Composition of decomposition gases.	Aerothermochemical properties of flowing gases in convection, diffusion, conduction and radiation; fuel input for combustion.	Pyrolysis followed by gas analysis (MS or GC).
3. Enthalpy of thermal decomposition of ablative solid.	Heat transfer.	Calorimetry, quantitative DTA, DSC. Evaluation requires composition of decomposition gases.
4. Char density	Conduction, vaporization, material transfer, strength of shield.	TG and sectioning of simulated ablation test samples.
5. Oxidation, combustion, melting, vaporization and sublimation of char and refractory reinforcements.	Material and energy transfer at surface.	Laboratory measurements of related phenomena, or scientific literature.
6. Thermal properties of virgin plastic and char.	Heat transfer.	DSC and special thermal conductivity apparatus.
7. Optical properties of surface.	Absorption and emission of radiation.	Special optical equipment.

315

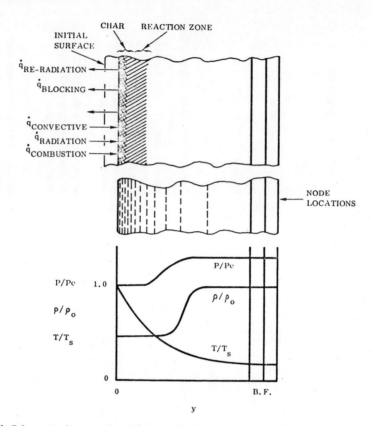

Fig. 1 Schematic diagram for ablation calculation program (3)

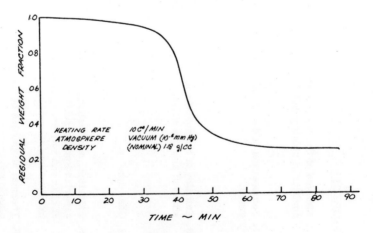

Fig. 2 TG curve of nylon-phenolic at 10° C/min (4)

316

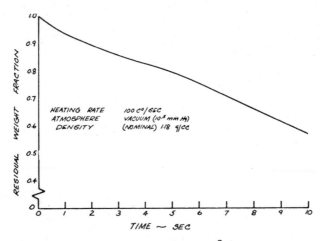

Fig. 3 TG curve of nylon-phenolic at 100° C/sec (4)

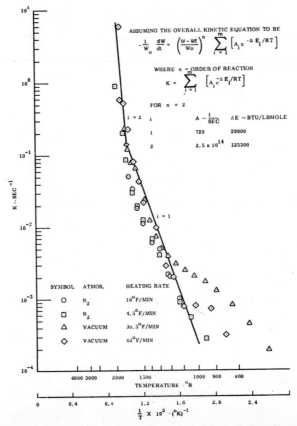

Fig. 4 Arrhenius plot of TG data for refractory reinforced phenolics (3)

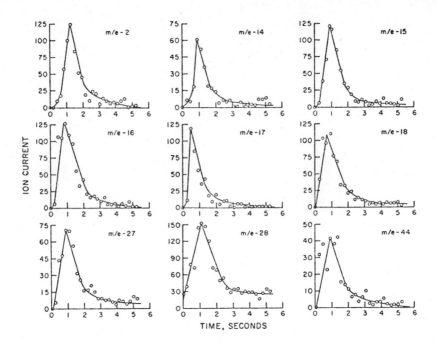

Fig. 5 Representative EGA-MS results from filament
heating of phenol-formaldehyde at 930° C/sec.
Part 1 (15)

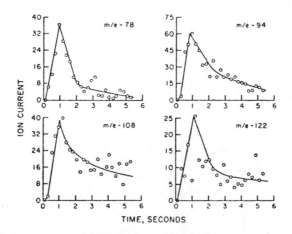

Fig. 6 Representative EGA-MS results from filament
heating of phenol-formaldehyde at 930° C/sec.
Part 2 (15)

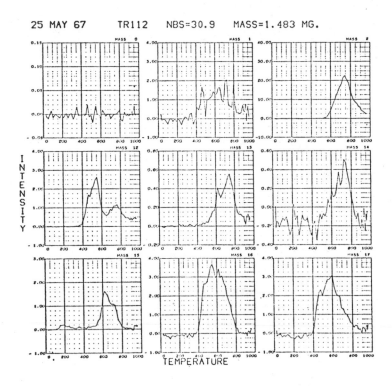

25 MAY 67   TR112   NBS=30.9   MASS=1.483 MG.

INTENSITY

TEMPERATURE

Fig. 7 EGA-MS data page for pyrolysis of
aromatic polyamide at 10° C/min. (19)

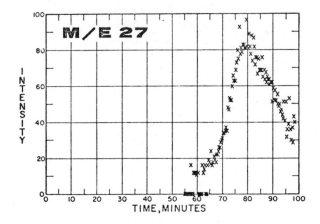

M/E 27

INTENSITY

TIME, MINUTES

Fig. 8 EGA-MS results for polyquinaxaline at m/e-27 (21)

319

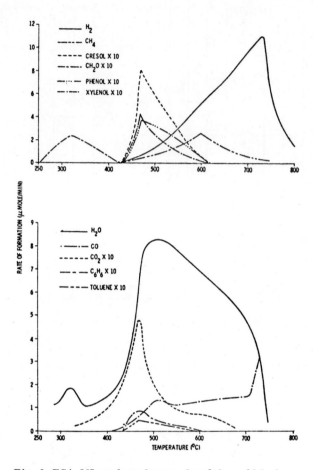

Fig. 9 EGA-MS products from a phenol-formaldehyde
        polycondensate (23)

320

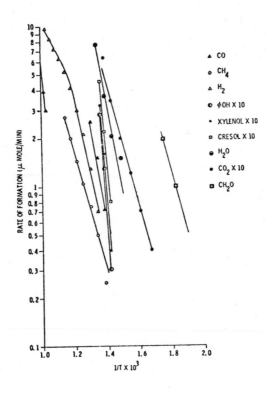

Fig. 10 Arrhenius plot for early formation of EGA–MS
products from a phenol-formaldehyde poly-
condensate (23)

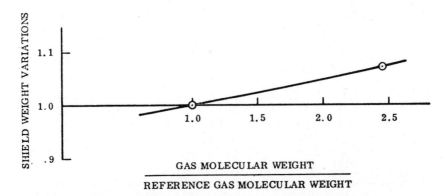

Fig. 11 Effect of uncertainty of molecular weight of injected gas species on
weight of nylon reinforced phenolic re-entry vehicle heat shield in
ballistic satellite trajectory (3)

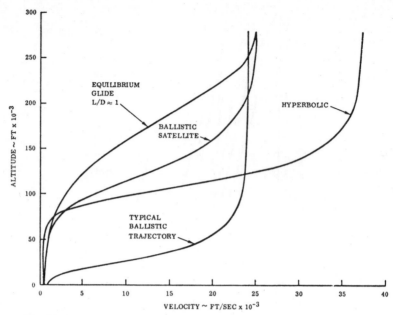

Fig. 12 Re-entry trajectories for missions analyzed by Shaw,
Garner and Florence (3)

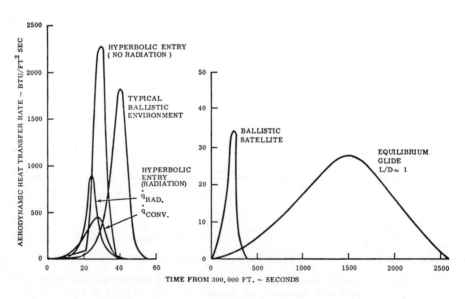

Fig. 13 Aerothermodynamic environments of missions analyzed by Shaw,
Garner and Florence (3)

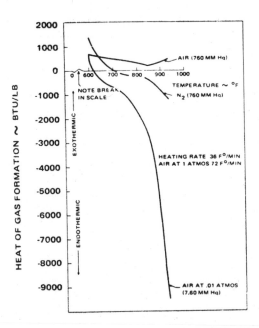

Fig. 14 Heat of gas formation of ESM 1030 in three different atmospheres

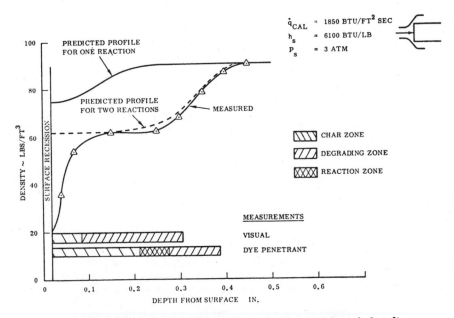

Fig. 15 Char density variation with depth for refractory reinforced phenolic
test sample heated in simulated re-entry environment (3)

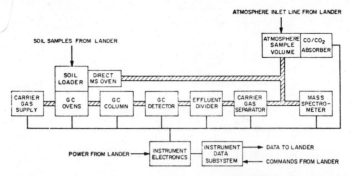

Fig. 16 Functional block diagram of 1976 Mars Viking Lander gas chromato-
graph-mass spectrometer system. Oblique parallel lines – gas flow
path; solid lines – power, control, and data circuits; GC–gas chromatograph; MS–mass spectrometer(28)

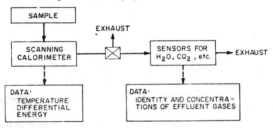

Fig. 17 Functional block diagram of proposed Mars Viking Lander soil water
apparatus. (28) Will not be included in 1976 mission.

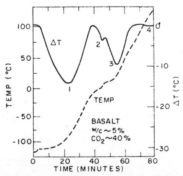

Fig. 18 Results of calibrated DTA on powdered basalt containing 40% solid
CO_2 and 5% solid H_2O mechanically mixed into the sample, in 15 mbar
CO_2 atmosphere. (1) corresponds to sublimation of solid CO_2;
(2) corresponds to fusion of ice; (3) corresponds to vaporization of
H_2O; (4) corresponds to desorption of H_2O. Presence of ice revealed
by ratio of (2) to (3) 1:7 (the ratio of the enthalpies of fusion and
vaporization) at 0^0 C and 7^0 C, the melting and boiling points at
the ambient pressure on Mars (28)

324

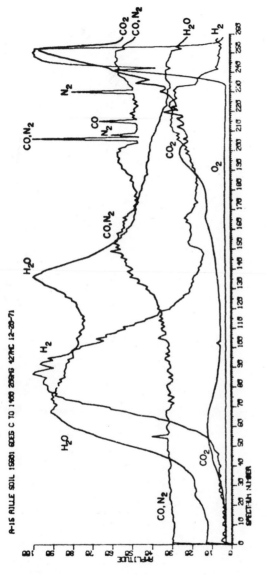

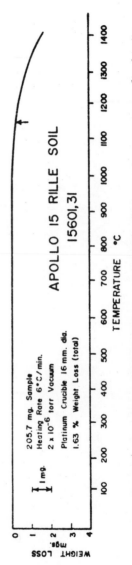

Fig. 19 EGA–MS and TG results from lunar soil sample. Each gas shown was normalized to its greatest release temperature (100% amplitude). H$_2$S and SO$_2$ were omitted. Arrow on TG curve shows initial melting of soil sample (36)